Problem Books in Mathematics

Series Editor

Peter Winkler, Department of Mathematics, Dartmouth College, Hanover, NH, USA

Books in this series are devoted exclusively to problems - challenging, difficult, but accessible problems. They are intended to help at all levels - in college, in graduate school, and in the profession. Arthur Engels "Problem-Solving Strategies" is good for elementary students and Richard Guys "Unsolved Problems in Number Theory" is the classical advanced prototype. The series also features a number of successful titles that prepare students for problem-solving competitions.

Adam Coffman • Justin Gash • Rick Gillman •
John Rickert

The Indiana College Mathematics Competition (2001–2023)

Celebrating the Teamwork Spirit
and the Peter Edson Trophy

Adam Coffman
Mathematical Sciences
Purdue University Fort Wayne
Fort Wayne, IN, USA

Justin Gash
Mathematics
Franklin College
Franklin, IN, USA

Rick Gillman
Mathematics and Statistics
Valparaiso University
Valparaiso, IN, USA

John Rickert
Mathematics
Rose–Hulman Institute of Technology
Terre Haute, IN, USA

ISSN 0941-3502   ISSN 2197-8506   (electronic)
Problem Books in Mathematics
ISBN 978-3-031-62767-5   ISBN 978-3-031-62768-2   (eBook)
https://doi.org/10.1007/978-3-031-62768-2

Mathematics Subject Classification: 00A07, 01A05, 97U40

© The Editor(s) (if applicable) and The Author(s), under exclusive license to Springer Nature Switzerland AG 2024

This work is subject to copyright. All rights are solely and exclusively licensed by the Publisher, whether the whole or part of the material is concerned, specifically the rights of translation, reprinting, reuse of illustrations, recitation, broadcasting, reproduction on microfilms or in any other physical way, and transmission or information storage and retrieval, electronic adaptation, computer software, or by similar or dissimilar methodology now known or hereafter developed.

The use of general descriptive names, registered names, trademarks, service marks, etc. in this publication does not imply, even in the absence of a specific statement, that such names are exempt from the relevant protective laws and regulations and therefore free for general use.

The publisher, the authors and the editors are safe to assume that the advice and information in this book are believed to be true and accurate at the date of publication. Neither the publisher nor the authors or the editors give a warranty, expressed or implied, with respect to the material contained herein or for any errors or omissions that may have been made. The publisher remains neutral with regard to jurisdictional claims in published maps and institutional affiliations.

This Springer imprint is published by the registered company Springer Nature Switzerland AG
The registered company address is: Gewerbestrasse 11, 6330 Cham, Switzerland

If disposing of this product, please recycle the paper.

Preface

Every springtime in Indiana since 1966, students have gathered from across the state to participate in the Indiana College Mathematics Competition.

At this event, also called "the Friendly Competition," everyone meets for a brief orientation, and then splits up into teams of three students, each getting two hours to work together to solve undergraduate-level math problems. Then everyone comes back together for dinner and discussion. The three top-scoring teams are announced the next day.

Because the ICMC is intended to be a student activity where all the participants can enjoy some success in problem solving, each year's problem set has both easier questions and more challenging ones, in a format where answers are to be written and explained, with the opportunity to get partial credit for a good start or the right general idea. The main topics are from the standard undergraduate curriculum for math majors:

Calculus	Combinatorics/Discrete Math
Linear Algebra	Probability
Number Theory	Geometry
Abstract Algebra	Real Analysis

Some problems are like textbook exercises or entrance exam questions for graduate study, while some others are elementary logic puzzles not using any advanced terminology or techniques. Some are original inventions by the test writers, and others are well-known, classic contest problems. This competition doesn't require Olympiad-level training and the memorization of obscure formulas or inequalities, just a good grasp of the knowledge from college math classes, problem-solving skills, and teamwork!

Readers interested in a specific topic can use the **Index** (starting on page 219) to find problems from a subject area (calculus, number theory, ...) or solutions using certain methods (induction, telescoping sums, ...). Each year's problem set appears as it did in the competition—with no subject labels, indication of difficulty, or hints

The student participants in the 2015 ICMC, held at Taylor University. Photo by Professor Jeremy Case, Taylor University [TU]

other than those occasionally included by the problem authors. The problems and solutions chapters are connected to each other by hyperlinks (in the electronic book) and page number references.

Fort Wayne, IN, USA   Adam Coffman
Franklin, IN, USA   Justin Gash
Valparaiso, IN, USA   Rick Gillman
Terre Haute, IN, USA   John Rickert

Historical Remarks and Acknowledgments

On April 27, 1965, Peter Edson, a trustee of Wabash College, sent its dean of faculty a memorandum that included a newspaper clipping about a unique mathematics competition that was held among high schools in New Jersey. In this competition, each school entered a team that worked as a team on a set mathematics examination. Edson wondered if anything of that sort was done at the college level. The dean, as deans do, handed the question off to one of his faculty members, Paul Mielke, who answered that he knew of no such competition but that he would be willing to suggest the idea to his Indiana colleagues. Response to his letter was immediate and favorable. Paul and others took up the challenge of creating such an exam, which has resulted in nearly 60 years of shared experiences for undergraduate mathematics students in the state of Indiana.

In this introduction, we briefly share the life stories of these two men and of the Indiana College Mathematics Competition which they inspired and brought to life.

Peter Edson

One might assume that Peter Edson was a well-known (at least at the time) mathematician intimately involved with college education in the State of Indiana. Surprisingly, while the second half of that sentence is accurate, the first is not. Peter Edson was a newspaper journalist, with no particular training in the mathematical sciences.

He was born in Hartford City, Indiana, in 1896 to parents who were teachers, with degrees from Hanover College, located in Madison, IN (Appendix). While he was still in high school, Peter worked as a correspondent for the *Fort Wayne News-Sentinel*. He was a bright, determined young man. In 1913, his high school newspaper wrote that "Pete was the pilot [of the Junior class]. He was the class grand problem solver and semi-official idea generator."

After graduating from high school and matriculating at Wabash College, he worked for the *Crawfordsville Journal* and the *Indianapolis Star*. He also served in

Peter Edson, Fort Wayne High School basketball uniform. Photo courtesy of the Edson family [ME$_{23}$]

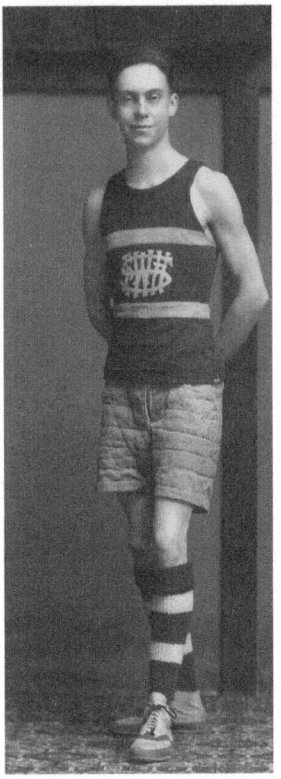

several editorial roles, including as Editor-in-Chief for the Wabash College student newspaper, *The Bachelor*. He was a young man who clearly seemed to know what he wanted to do with his life!

However, Peter's life took a detour in 1917. That year he left Wabash College to enter Reserve Officer Training prior the United States' entry into WWI. Peter served as a Lieutenant in the infantry until he was discharged in 1919. After the war, Edson returned to Wabash College and his service on *The Bachelor*. A member of Phi Beta Kappa, he graduated in 1920, worked briefly as a roustabout in the Ohio oilfields, and then began working as the Sunday Editor at the *Fort Wayne News-Sentinel*.

This was the beginning of a life-long career as a professional journalist. Peter earned a Master of Arts degree in 1925 from Harvard University. Concurrently, he joined the *Boston Post*. From there, he moved to the *New Haven Register* and then on to the *Pittsburgh Press*. By 1927 he was the editor for *Every Week Magazine*. In 1932 he was appointed as the Editor-in-Chief of the Newspaper Enterprise Association (NEA), a syndication service.

In 1941, Edson was promoted by the NEA to be its lead Washington correspondent, a post he held until his retirement in 1964. In the archives of his papers, held by the Wisconsin Historical Society [WHS], there are many congratulatory messages

Historical Remarks and Acknowledgments

Peter Edson, Wabash fraternity photo (at right, in white shirt). Photo courtesy of the Edson family [ME$_{23}$]

Peter Edson's desk in Washington. Photo courtesy of the Edson family [ME$_{23}$]

on this promotion and praise for his work as Editor-in-Chief. As he stepped into this new position, he wrote that his vision was to write "a very informal piece, going after the color and humor and the human interest of Washington. There is a gold mine of this copy which no one seems to have staked a claim on, and I'd like to make that mine, mine."

An early example of Peter's careful, and popular, writing style is a 1941 column about the Lend-Lease program, just passed by Congress, titled "How much is $1.3 billion? Breakdown of Lease-Lend Limit Shows Big Sum." In the column, Edson carefully explained what could be purchased with this amount of money. He begins by noting that $500,000,000 goes, off the top, to Winston Churchill's request for "war tools." Churchill also asked for 50 destroyers, costing $8,000,000 each for another $400,000,000, leaving the remaining $400,000,000 for planes and other ships.

Edson reports that the British were losing 68,000 tons of shipping per week and could only replace half of it themselves. At the time, cargo ships were averaging $333 per ton to build, but a recent contract demonstrated that this could be brought down to $166 per ton. At the lower price, the US could build the needed tonnage for a total of approximately $295,000,000. Then, given a price of $125,000 per patrol bomber, we can supply 850 planes to the British with the remaining $105,000,000.

Edson reminds readers that the British could swap eight destroyers for a battle ship, or 200 planes for a cruiser. And finally, he wryly notes that if you did not like these numbers, then there is always next year's budget.

Similarly, a 1942 column on the problems of the Japanese-American internment program foreshadowed a dark chapter in American history, and a 1945 column on post-war immigration laid out the pros and cons to generous immigration policies following the war before coming down decisively in favor of immigration.

His work, and work ethic, quickly made Peter a nationally recognized journalist. He won several journalism awards including the 1945 Sigma Delta Chi Award for Journalism and the 1952 Headliners Award for Reporting. In 1948 he was awarded the Raymond Clapper Memorial Award for excellence in Washington Reporting. This award was identified by the White House Correspondents Association (WHCA) as its highest award after Clapper's death in 1944. The Award was given by the WHCA to a journalist or team for distinguished Washington reporting.

Peter Edson's most notable journalistic achievement was his writing that broke the Nixon slush fund scandal which led to Nixon's now famous "Checkers" speech. By the mid-1950s, his column was being carried in over 800 newspapers and being read by more than 13 million people weekly. In a report on the reach of the NEA, an association vice-president wrote that "this should make Edson far and away the most widely printed Washington columnist."

Peter was a modest man who was a firm believer in teamwork and in sharing credit appropriately. For example, The Raymond Clapper Award came with a cash prize which Peter promptly re-distributed to his more junior colleagues.

But how did Peter Edson get connected to an Indiana mathematics competition? Clearly, he was a Hoosier. As noted above, he had heard of a team mathematics competition for high school students in New Jersey. Peter asked the dean of the faculty if there were anything like this in Indiana. And the rest is history, as one might say.

Peter's daughters also shared that he had designed the Washington DC home, both the original home and an addition, that they grew up in. They also told the story of Edson's ability to modify a self-closing gate they saw in Williamsburg, VA,

Historical Remarks and Acknowledgments   xi

for their home garden, and his fascination with the organization and efficiency of the Disney World they visited on a family vacation [ME$_{22}$, MA].

Edson was deeply attached to his home state of Indiana. He returned frequently for visits and never lost his Hoosier values of modesty and hard work. He served on the Wabash College Board of Trustees from 1952 to 1964. The college awarded him the 1955 Alumni Award of Merit and, in 1971, an honorary Doctor of Humane Letters.

Peter died in 1977.

Paul Mielke

Paul Mielke was born in Racine, Wisconsin, in 1920, the only son in a family with five daughters. In contrast to Peter Edson, his mathematical aptitude showed at an early age while he spent time working with his father, who was an expert tool and die maker. Paul was also an expert carpenter, personally creating the trophies awarded for the competition during its early years. In addition to his skills with tools, Paul was a very talented draftsman, producing beautiful drawings using India ink with the stainless steel quills in his drafting set.

Paul eventually matriculated to Wabash College, where he earned an A.B. in mathematics in 1942. Service in the Marine Corps during WWII interrupted his study at Brown University, where he eventually, in 1946, earned a Sc. M. in Applied Mathematics. He returned to Indiana, teaching briefly at Wabash College before earning his Ph.D. in Mathematics at Purdue University in 1951.

Although his thesis was in mathematical analysis [M$_1$] and much of his work was in that area, Paul had a love of the classical subjects of number theory and geometry, presumably drawing on the craftsmanship of his training as a child. One interesting intersection of Paul's practical and mathematical interests was in compass and straightedge constructions. The one that impressed his son the most was that he could construct three-dimensional paper models of the five Platonic solids using nothing more than a T square, a compass, a piece of heavy drafting paper, a pair of scissors, and some Elmer's glue [WCNews].

Mielke returned to Wabash College after completing his degree at Purdue. He continued to work there throughout his professional career. While advancing in faculty rank, Paul also founded and served as head of the college's computer lab in the early 1960s and also served as chair of the Department of Mathematics from 1963 to 1978. The college awarded him an honorary degree a few years after his retirement in 1985.

As many mathematicians do, Paul took leaves from Wabash to engage in related professional work. The years between 1952 and 1957 saw him in Seattle working for Boeing Airline Company as a structural dynamics engineer. The opportunity was attractive for a number of reasons: it gave him the chance to be on the leading edge of applying the new technology of computers to solve real world problems; the Korean War, which he feared would lead to WWIII, had just started and this was a

This photograph of Paul Mielke appears here courtesy of the Robert T. Ramsay Jr. Archival Center, Wabash College, Crawfordsville, IN [WCLib]

chance to contribute to the national defense. The Boeing team built one of the first industrial computer labs and used the machines to solve differential equations.

The year 1969 found him on leave to work as the director of the Mathematical Association of America's Committee on the Undergraduate Program in Mathematics, the body which sets the standards for the undergraduate curriculum. He was active in the Association throughout his career, holding many leadership roles in the Indiana Section of the MAA and being recognized for this work by receiving the section's first Distinguished Service Award. Besides leading the CUPM, Paul also served on the Association's Board of Directors and was an associate editor for the *American Mathematical Monthly* from 1974 to 1978.

Paul published sparingly, but in quality publications, with articles in the *Mathematics Teacher* [M_2] and in the *Two Year College Mathematics Journal* (now the *College Mathematics Journal*) [M_3]. However, his enthusiasm for teaching was palpable and it is evident that many of his students felt the depth of his commitment to teaching.

He was always prepared. For every textbook, he had completely worked out the answer to every problem in every chapter. As Paul's son relates, "He took special satisfaction in the design of the new classrooms in Baxter with chalk boards on two or three walls of the room and chairs that rotated in place. I recall several lectures in which he had his presentation carefully planned so that he used all the available chalkboard space and didn't have to erase anything. He would complete the peroration of his lecture and punctuate the conclusion with his chalk as he completed the last proof at the bottom of the last available panel of the chalkboard." [WCNews]

Historical Remarks and Acknowledgments xiii

This photograph of Paul Mielke appears here courtesy of the Robert T. Ramsay Jr. Archival Center, Wabash College, Crawfordsville, IN [WCLib]

If Peter Edson was a journalist with hidden mathematical talents, Paul was a mathematician with other hidden talents and interests. Besides the woodworking skills mentioned above, his daughter wrote that her dad loved dancing and photography, winning local prizes in both areas. More significantly, he was active in the Civil Rights movement of the 1960s, founding and leading the Crawfordsville, IN, branch of the NAACP as it raised funds and donations for the movement.

Paul passed away in 2008.

The Indiana College Mathematics Competition

The convergence of Peter Edson's and Paul Mielke's lives and interests is not quite as surprising as it might seem. Both had life-long love of the state of Indiana, and particularly of Wabash College which they attended and later served in some capacity. Both men valued college education and the need to challenge young men and women to excel. Further, they understood the value of teamwork and of rewarding success.

The Indiana College Mathematics Competition (ICMC) was launched in 1965 by the Indiana Section of the Mathematical Association under Mielke's leadership. As a small college competition (the competition was originally restricted to students at the private colleges in Indiana), it was intended to serve as counterpoint to the nationally distributed but notoriously difficult Putnam Exam. To this point, the exam focused on mathematical topics in the undergraduate curriculum and

This photograph of Paul Mielke appears here courtesy of the Robert T. Ramsay Jr. Archival Center, Wabash College, Crawfordsville, IN [WCLib]

was intentionally designed to be completed by a team of students, rather than by individuals. It retains these characteristics today. A brief history of the early years of the ICMC written by Professor Mielke himself appears in [AFMC].

Typically, the ICMC is held on the first afternoon of the spring meeting of the Indiana Section of the Mathematical Association of America (MAA). Student teams check-in during the mid-afternoon after driving to the meeting site. After general instructions are given, the teams are dispersed to their assigned work spaces, often with students from the host institution showing them around campus. This workspace is typically a classroom, but as the number of teams increases they are often assigned to conference rooms—any place with a blackboard or a whiteboard! They then have two hours to complete the exam and return it, all in time to attend the conference banquet and listen to the keynote speaker.

A team of faculty who have volunteered to be graders work overnight to mark the exams on a scale of 0–10 for each problem, with partial credit awarded according to the solution guides from the problem writers and the impressions of the graders. Winners are announced at the following day's business meeting, held after morning sessions which often include a workshop specifically designed for undergraduate students.

The winning team is awarded the Peter Edson Trophy. Before his death, Paul Mielke would handcraft this trophy from native Indiana wood.

The 2003 trophy for the first place team from Indiana–Purdue Fort Wayne

The ICMC's strict two-hour limit, including walking time, puts an interesting limit on the complexity and obscurity of the problems. Each individual exam reflects the idiosyncrasies of the writers, and the difficulty of the problem sets can vary significantly from year to year. In preparing for and participating in competitions like the ICMC, students often learn about mathematics not covered in their college courses. It seems unlikely that a single student would have taken courses in all of the topics covered by the exam, implying that the "team" element of the competition is essential. However, when viewed over the long-term history of the exam, the problems reflect an evolving core of mathematical knowledge in the undergraduate curriculum.

In the spring of 2020, the Indiana Section meeting, including the ICMC, was abruptly but necessarily canceled, under the circumstances of the onset of the COVID-19 pandemic. The 2021 ICMC event was temporarily re-organized into a simultaneous, but remote, locally proctored format. In 2022, the competition returned to its in-person format.

Acknowledging the Volunteers Who Organize the ICMC

The Mathematical Association of America, MAA, is a national professional organization founded in 1915 for college mathematics teachers and students, with over 25,000 faculty and student members, and a large budget and professional full-time staff. The operations of each of its 29 regional Sections, including the Indiana Section which has been active since 1924, are organized by volunteer members with a small budget. The Indiana College Mathematics Competition has been the main student activity at meetings of the Indiana Section since 1966; the competition was quickly nicknamed the "Friendly" Competition because of its focus on solving mathematical problems with teamwork, and bringing faculty and students together from around the state, rather than a focus on Olympiad-level individual competition.

We (the editors of this volume: AC, JG, RG, JR) would like to recognize the work of many individuals who have contributed to the success of the Indiana College Mathematics Competition over the past couple of decades.

These individuals include the mathematics professors who have served as the Indiana Section's Student Activities Coordinator each year, leading the organizing of the ICMC event: Aaron Klebanoff (2000–2001), Mohammad Azarian (2001–2007), Robert Talbert (2007–2010), Justin Gash (2010–2016), Paul Fonstad (2016–2019), and Colin McKinney (2019–2023). The Coordinator is an officer on the Executive Board of the Indiana Section of the MAA, and all the Board members contribute to organizing the Section meeting at a different host institution every year, from among the colleges and universities in Indiana (listed in Appendix).

Each year, the exam and a solution set are assembled by a committee from the local host institution, or by an individual or team recruited by the Student Activities Coordinator. The writers compile a list of problems, guided only by their motivation to provide a fun and interesting student activity. In particular, the contest writers have not been obliged to invent new, original problems, but have sometimes copied or adapted problems from past ICMC events or similar contests, or drawn from the math problem folklore. We would like to thank all these writers for their creative contributions to the success of mathematics education and the mathematics community in Indiana.

The Student Activities Coordinator also recruits graders, usually a group of faculty attending the Spring Meeting, with each grader evaluating the teams' answers to one or two problems and reporting their scores the morning after the competition. The host institutions and local organizers of the spring Section meetings all deserve acknowledgment. They provided the competitors with efficient registration procedures, excellent campus guides, and many classrooms so that each team could work in their own space. Without their help, this competition would not run as smoothly as it has for the past 20 years.

And, finally, we would like to thank the many dedicated coaches and hundreds of undergraduate students who have participated in the competition. They are the folks who put the "friendly" in the name of the competition and bring joy to their home institutions with their interest and enthusiasm in mathematical problem solving.

About This Volume

The problems and solutions from the first 35 years (1966–2000) of the competition were gathered in the book *A Friendly Mathematics Competition* [AFMC]. Chapter 47 of this work contains some corrections to that volume as well as new solutions to some of the older problems. The problems in this collection that are repetitions or variations of these older problems from [AFMC] are indicated in remarks in the solution sets. Some of the problems here adapted from other sources are also

Historical Remarks and Acknowledgments xvii

marked with a link to the list of references, which gives the reader a sample of problem books at a level similar to the ICMC [GKL, HWi, K_1, K_2, RL].

In this compilation (2001–2023), all the problems are copied as presented in the contests, although a few statements have been edited for clarity or correctness, in the interest of prioritizing usefulness to the reader over historical preservation. As the solution sets provided by the exam authors at the time of the contests were intended only as guides for the graders, some of those solutions were sketches for experts rather than detailed, model solutions for students. So, many of the solutions in Chaps. 24–45 are expanded versions of the originals distributed after each competition. Some expanded, re-written, or alternate solutions are based on the real-time answers from student contestants, or on feedback from reviewers well after the contest, and some solutions are entirely new. For some problems, writing a solution with the amount of detail appearing here may not be practical or strategic for students in a two-hour contest situation. However, the ICMC exams are designed to allow for partial credit to be assigned by the grader on a 10-point scale for each problem, so a submitted solution judged to have some correct steps, a good partial solution, or the right general idea can get some points.

The editors are pleased to thank the past and present members of the Executive Board of the Indiana MAA for their support. In particular, Professor Vesna Kilibarda of Indiana University Northwest was Indiana Section Chair in 2010–2011, and she was an original member of the committee exploring the possibility of a sequel to [AFMC].

We thank the competition's most prolific contributors, Mike Axtell and Joe Stickles, for sharing their original electronic files, which filled some gaps in our records. The editors also thank Professor Steve Kennedy for his support during early stages of this project, and Professor George Gilbert and several anonymous reviewers, whose generous suggestions improved the content and contributed some of the alternate solutions.

Credit for some photos appearing here is given in the appendix list of references—photos appearing without credit were shared here by A. Coffman.

The historical sections on Edson and Mielke were prepared by R. Gillman, in correspondence with members of the Edson family [ME_{22}, ME_{23}, MA], and with the assistance of Elizabeth Swift, the archivist at Wabash College [WCLib].

Any author royalties or payments from sales of this volume will be donated to the Indiana Section of the MAA to support future competitions and other student-focused activities.

Contents

Part I Problems

1	2001 Problems	3
2	2002 Problems	5
3	2003 Problems	7
4	2004 Problems	9
5	2005 Problems	11
6	2006 Problems	13
7	2007 Problems	15
8	2008 Problems	17
9	2009 Problems	19
10	2010 Problems	21
11	2011 Problems	23
12	2012 Problems	27
13	2013 Problems	29
14	2014 Problems	31
15	2015 Problems	35
16	2016 Problems	39
17	2017 Problems	45
18	2018 Problems	49
19	2019 Problems	51
20	2020 Problems	55

21	2021 Problems	57
22	2022 Problems	59
23	2023 Problems	61

Part II Solutions

24	2001 Solutions	65
25	2002 Solutions	71
26	2003 Solutions	75
27	2004 Solutions	81
28	2005 Solutions	85
29	2006 Solutions	91
30	2007 Solutions	95
31	2008 Solutions	99
32	2009 Solutions	103
33	2010 Solutions	109
34	2011 Solutions	113
35	2012 Solutions	121
36	2013 Solutions	131
37	2014 Solutions	137
38	2015 Solutions	145
39	2016 Solutions	151
40	2017 Solutions	159
41	2018 Solutions	167
42	2019 Solutions	173
43	2021 Solutions	181
44	2022 Solutions	187
45	2023 Solutions	193

Part III More History of the ICMC

46	Top Scoring Teams	201

47	**Updates to *A Friendly Mathematics Competition***	207
	Some Errata for the 1966–2000 Problem Book	207
	New Solutions for Old Problems...	208

Appendix 213
    Location Index 213
    References and Photo Credits 216

Index.. 219

Part I
Problems

Chapter 1
2001 Problems

P2001-1 Suppose that n is a positive integer, not a power of 2. Prove that n is the sum of two or more consecutive positive integers.

P2001-2 Let $f(x) = x^{1/x}$, $x > 0$. Prove that f has exactly two inflection points. Hint: $f''(x) = \dfrac{x^{1/x}}{x^4}\left((\ln(x))^2 + (2x-2)\ln(x) + (1-3x)\right)$.

P2001-3 A machine is required to drop k indistinguishable balls into n numbered slots, $k \geq n$, and must drop at least one ball into each slot. How many outcomes may arise?

P2001-4 There is a cubic polynomial f with integer coefficients and the property that $f(\cos(\theta)) = \cos(3\theta)$:

(a) Find f.
(b) Prove that $\cos\dfrac{\pi}{9}$ is irrational.

P2001-5 A circle C_1 of radius 1 is inside of and tangent to a circle C_2 of radius 2. C_1 rolls around the interior of C_2 without slipping, making continuous tangential contact with C_2, and returns to its initial position. Describe the locus of points traced out by a distinguished point P on the circumference of C_1.

P2001-6 Is $\displaystyle\sum_{n=1}^{\infty} \dfrac{|\sin(n)|}{n}$ finite or infinite?

The 2001 Problem Solutions begin on page 65 in Chap. 24.

Chapter 2
2002 Problems

P2002-1 Consider the expression

$$\frac{f(x)}{f(x) + f(b-x)}.$$

Note that depending on f, it is possible for the expression to be undefined for some or all x in the interval $[0, b]$, so take care with the domains of the expressions appearing in the following parts (a) and (b):

(a) Find a function $f(x)$ which is continuous on $[0, b]$ and not identically zero, such that $f(x) + f(b-x) = 0$ for all x in the interval $[0, b]$.
(b) For f continuous and positive on $[0, b]$, evaluate

$$\int_0^b \frac{f(x)}{f(x) + f(b-x)} dx.$$

P2002-2 Let $\sum a_n$ be a series of strictly positive terms, and let

$$b_n = \frac{a_1 + a_2 + \cdots + a_n}{n}.$$

Show $\sum b_n$ diverges.

P2002-3 Show that the sum of two consecutive odd primes has at least three (not necessarily distinct) prime factors.

P2002-4 A university bookstore sold at least one mathematics textbook each day for 100 consecutive days. During this time, the bookstore sold 140 mathematics textbooks. Was there a period of consecutive days when exactly 59 mathematics textbooks were sold?

P2002-5 Let φ be a function from a set S with a binary operation (denoted by juxtaposition) to the set of nonnegative integers such that $\varphi(xy) = \varphi(x)\varphi(y)$ for all x and y in S:

(a) Supposing S is a group and the binary operation is the group operation, show that either $\varphi(x) = 0$ for all x in S, or $\varphi(x) = 1$ for all x in S.
(b) Supposing φ is a nonconstant function, S is a ring, and the binary operation is the ring multiplication, show that $\varphi(0_S) = 0$.

P2002-6 Show that a square inscribed in a triangle can enclose at most one half the area of the triangle.

The 2002 Problem Solutions begin on page 71 in Chap. 25.

Chapter 3
2003 Problems

P2003-1 Let S be a set with a binary operation $*$ that has the property $(a * b) * (c * d) = a * d$ for all $a, b, c, d \in S$. Prove these statements:

(a) If $a * b = c$, then $c * c = c$.
(b) If $a * b = c$, then $a * y = c * y$ for all $y \in S$.

P2003-2 A tennis club invites 2^k ($k \geq 1$) players of equal ability to compete in a single-elimination tournament. Suppose that the probability of a competitor winning his or her match is $\frac{1}{2}$, and that the competitors are paired together in a random order in the first round. Show that the probability that a particular pair of players will compete against each other in this tournament is $\dfrac{1}{2^{k-1}}$.

P2003-3 Five points divide the unit circle into five equal arcs. Express ab in terms of $\sin 18°$, where a and b are the lengths of the chords indicated in the picture here:

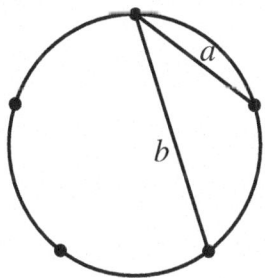

P2003-4 A 3×3 array of integers has the property that the products of the elements in each of the three rows, in each of the three columns, and in each of the two diagonals are all the same; call this product k. (We call such an array a *multiplicative magic square*.) Prove that k is a perfect cube, and determine the element in the array of which k is the cube.

P2003-5 Does the series

$$1 + \frac{1}{2} - \frac{1}{3} + \frac{1}{4} + \frac{1}{5} - \frac{1}{6} + \frac{1}{7} + \frac{1}{8} - \frac{1}{9} \cdots$$

converge or diverge? Justify.

P2003-6 The Fibonacci sequence $f_1, f_2, f_3, \ldots$ is defined by $f_1 = f_2 = 1$,

$$f_n = f_{n-1} + f_{n-2}$$

for $n = 3, 4, 5, \ldots$. Thus, the sequence begins

$$1, 1, 2, 3, 5, 8, 13, 21, 34, \ldots.$$

Let

$$Q = \begin{bmatrix} 1 & 1 \\ 1 & 0 \end{bmatrix}:$$

(a) Prove

$$Q^n = \begin{bmatrix} f_{n+1} & f_n \\ f_n & f_{n-1} \end{bmatrix}$$

for $n = 2, 3, 4, \ldots$.

(b) Establish the identity:

$$f_{3n} = f_{n+1}^3 + f_n^3 - f_{n-1}^3$$

for $n = 2, 3, 4, \ldots$.

The 2003 Problem Solutions begin on page 75 in Chap. 26.

Chapter 4
2004 Problems

P2004-1 Partition the set $\{1, 2, 3, 4, 5\}$ into two arbitrarily chosen sets. Prove that one of the sets contains two numbers and their difference.

P2004-2 Suppose $a > 1$:

(a) Show the series $\sum_{n=0}^{\infty} \dfrac{2^n}{a^{2^n} + 1}$ converges.

(b) Determine to what value this series converges.

(Note that a correct solution for part (b) necessarily solves part (a) as well.)

P2004-3 Let A be a 4×4 matrix such that each entry of A is either 2 or -1. Let $d = \det(A)$; clearly d is an integer. Show that d is divisible by 27.

P2004-4 Let $a_1, a_2, \ldots, a_n$ be a finite sequence of real numbers. Form a sequence of length $n - 1$ by averaging two consecutive terms of the sequence:

$$\frac{a_1 + a_2}{2}, \frac{a_2 + a_3}{2}, \ldots, \frac{a_{n-2} + a_{n-1}}{2}, \frac{a_{n-1} + a_n}{2}.$$

Continue this process of averaging two consecutive terms until you have only one term left. Show that this final term is

$$\frac{\sum_{i=0}^{n-1} \binom{n-1}{i} a_{i+1}}{2^{n-1}}.$$

P2004-5 Let P be the center of a square with side $\overline{AC}$. Let B be a point in the exterior of the square such that $\triangle ABC$ is a right triangle with hypotenuse $\overline{AC}$. Prove: $\overline{BP}$ bisects $\angle ABC$.

P2004-6 Two ferryboats start at the same instant from opposite sides of a river, traveling across the water on routes at right angles to the parallel shores. Each travels at a constant speed, but one is faster than the other. They pass at a point 720 yards from the nearest shore. Both boats remain at their slips 10 min before starting back. On their return trips, they meet 400 yards from the nearest shore. How wide is the river?

The 2004 Problem Solutions begin on page 81 in Chap. 27.

Chapter 5
2005 Problems

P2005-1 Evaluate the integral

$$\int_{-1}^{1} \frac{|\sin(n \cos^{-1}(x))|}{\sqrt{1-x^2}} dx.$$

P2005-2 All points in the plane are colored in red, white, or blue. Prove that there is at least one pair of points of the same color with distance between them one unit.

P2005-3 Find the limit of the sequence defined by

$$a_n = \frac{1}{n^3} \sum_{k=1}^{n} \ln(1+kn).$$

P2005-4 Let $\mathbb{Q}^+$ be the set of all positive rational numbers and let "$*$" be an operation on $\mathbb{Q}^+$ that satisfies the following conditions for all $a, b, c, d \in \mathbb{Q}^+$:

$$(a * b)(c * d) = (ac) * (bd),$$
$$a * a = 1,$$
$$a * 1 = a.$$

Compute the value of the expression $((6/5) * (8/15)) * 2$.

P2005-5 Consider matrices A and B in $M_n(\mathbb{R})$ such that $A^3 = A^2$ and $A+B = I_n$. Show that the matrix $AB + I_n$ is non-singular and find its inverse.

P2005-6 Determine the real constants a, b, c, and p, such that

$$\lim_{x\to\infty}\left[\sqrt{9x^4 - 24x^3 + 6x^2 + 5} - (ax^p + bx + c)\right] = \frac{7}{3}.$$

P2005-7 Let M and N be the midpoints of BC and CD in the parallelogram $ABCD$, and let P be the intersection of AM and BN. Determine the ratios $\dfrac{AP}{AM}$ and $\dfrac{BP}{BN}$.

P2005-8 Given the integers x, y, and z, prove that if 25 divides the sum $x^5+y^5+z^5$, then 25 divides at least one of the numbers $x^5 + y^5$, $x^5 + z^5$, or $y^5 + z^5$.

The 2005 Problem Solutions begin on page 85 in Chap. 28.

Chapter 6
2006 Problems

P2006-1 Find the maximum value of the function

$$f(x) = \sqrt{4x - x^2 + 12} - \sqrt{10x - x^2 - 24}.$$

P2006-2 Evaluate

$$\sum_{n=1}^{\infty} \arctan\left(\frac{2}{n^2}\right).$$

Hint: It may be helpful to remember that

$$\arctan(\alpha) - \arctan(\beta) = \arctan\left(\frac{\alpha - \beta}{1 + \alpha\beta}\right).$$

P2006-3 Let X be a continuous random variable having the probability density function

$$f(x) = \begin{cases} \frac{1}{x^2}, & x \geq 1 \\ 0, & \text{otherwise.} \end{cases}$$

Find $P\left(\left[\sqrt{10X}\right] = 20 \,\Big|\, \left[\sqrt[3]{X}\right] = 3\right)$ (where $[x]$ denotes the greatest integer less than or equal to x).

P2006-4 A bicyclist rides 18 miles in exactly 72 min. Assuming the bicyclist's position is a continuous, increasing function of time, prove that there exists a contiguous 3-mile segment within this 18 miles that the rider completed in exactly 12 min.

P2006-5 Let S be a set of real numbers that is closed under multiplication. Let T and U be disjoint subsets of S whose union is S. Show that if the product of any three (not necessarily distinct) elements of T is in T and the product of any three elements of U is in U, then at least one of the two subsets T and U is closed under multiplication.

P2006-6 Let r and s be nonzero integers. Prove that the equation

$$(r^2 - s^2)x^2 - 4rsxy - (r^2 - s^2)y^2 = 1$$

has no solutions in integers x and y.

P2006-7 Equilateral triangle ABC, with points P on side $\overline{AB}$ and Q on side $\overline{AC}$, has been creased and folded along $\overline{PQ}$ so that vertex A lies at the point A' on $\overline{BC}$. Assuming $BA' = 1$ and $A'C = 2$, find the length of $\overline{PQ}$.

P2006-8 An evil genie pops out of a lamp and presents you with twelve identical-looking stones. The genie tells you that one stone has an imperfection not visible to the naked eye that causes the stone to be either slightly heavier or slightly lighter than the others. The genie also presents you with a balance scale and says you may use this scale at most three times. The genie will give you your heart's desire if you can correctly identify the imperfect stone and determine whether it is heavier or lighter than the others.

The 2006 Problem Solutions begin on page 91 in Chap. 29.

Chapter 7
2007 Problems

P2007-1 Let p and q be distinct primes. Find a nonconstant polynomial with integer coefficients that has $\sqrt{p} + \sqrt{q}$ as a root.

P2007-2 What is the value of the positive integer n for which the least common multiple of 36 and n is 500 greater than the greatest common divisor of 36 and n?

P2007-3 Evaluate: $\lim\limits_{x \to \infty} (x+2) \cdot \int_x^{3x} \dfrac{dt}{t\sqrt{t^4+1}}$.

P2007-4 Answer the following questions about prime numbers:

(a) Let p be a fixed prime and N a positive integer. Suppose an integer a is selected at random from the set $\{1, 2, 3, \ldots, Np\}$. What is the probability that a is divisible by p? (Think about the possible remainders when dividing by p.)

(b) For p, N as in (a), suppose two integers a and b are selected at random (independently, with replacement) from $\{1, 2, 3, \ldots, Np\}$. What is the probability that a and b are both divisible by p?

(c) Let $\{2, 3, 5, 7, \ldots, p_k\}$ be the list of the first k prime numbers, and denote their product $K = 2 \cdot 3 \cdot 5 \cdots p_k$. Let $P(k)$ denote the probability that two integers a and b, selected randomly from the list $\{1, 2, 3, \ldots, K\}$, have no common prime factor from the list $\{2, 3, 5, \ldots, p_k\}$. Assuming the limits exist, show $\lim\limits_{k \to \infty} P(k) = \prod\limits_{p \in \mathcal{P}} \left(1 - \dfrac{1}{p^2}\right)$, where $\mathcal{P}$ is the set of all primes.

P2007-5 Let A be an $n \times n$ matrix such that $a_{ij} = 1$ when $i \neq j$, and $a_{ij} = 0$ when $i = j$. In other words, $A = \begin{bmatrix} 0 & 1 & 1 & \cdots & 1 \\ 1 & 0 & 1 & \cdots & 1 \\ 1 & 1 & 0 & \cdots & 1 \\ \vdots & \vdots & \vdots & \ddots & \vdots \\ 1 & 1 & 1 & \cdots & 0 \end{bmatrix}$. Find A^{-1}. Hint: Using the matrix

$B = \begin{bmatrix} 1 & 1 & 1 & \cdots & 1 \\ 1 & 1 & 1 & \cdots & 1 \\ 1 & 1 & 1 & \cdots & 1 \\ \vdots & \vdots & \vdots & \ddots & \vdots \\ 1 & 1 & 1 & \cdots & 1 \end{bmatrix}$ may be helpful.

P2007-6 Let g and h be non-commuting elements in a group of odd order. Assuming g and h satisfy the relations $g^3 = e$ and $ghg^{-1} = h^3$, determine the order of h.

The 2007 Problem Solutions begin on page 95 in Chap. 30.

Chapter 8
2008 Problems

P2008-1 Let S be a set with a binary operation $*$ that is associative. Suppose that for all x and y in S we have $x * x * x = x$ (i.e. $x^3 = x$) and $x * x * y = y * x * x$ (i.e. $x^2 y = yx^2$). Show that for all x and y in S we have that $x * y = y * x$.

P2008-2 Two friends agree to meet at the library, but each has forgotten the time they were supposed to meet. Each remembers that they were supposed to meet sometime between 1:00 pm and 5:00 pm. They each independently decide to go to the library at a random time between 1:00 pm and 5:00 pm, wait for 30 min, and leave if the other does not show up. What is the probability that they meet during this 4-h period?

P2008-3 Suppose that two triangles have a common angle. Show that the sum of the sines of the angles will be larger in that triangle where the difference of the remaining two angles is smaller. Hint:

$$2\sin\left(\frac{\theta + \delta}{2}\right) \cos\left(\frac{\theta - \delta}{2}\right) = \sin(\theta) + \sin(\delta).$$

P2008-4 The base of a solid object is the region bounded by the parabola $y = \frac{1}{2}x^2$ and the line $y = 2$; cross-sections of the object perpendicular to the y-axis are semicircles. What is the volume of the object?

P2008-5 Define the sequence $\{x_n\}_{n=0}^{\infty}$ by $x_0 = 0$, $x_1 = 1$,

$$x_n = \frac{x_{n-1} + (n-1)x_{n-2}}{n}.$$

Determine $\lim_{n \to \infty} x_n$. Hint: Recalling that

$$\ln(1+x) = \sum_{k=0}^{\infty} \frac{(-1)^k}{k+1} x^{k+1}, \quad x \in (-1, 1]$$

may be helpful.

P2008-6 Prove that $\dfrac{1}{n+1}\dbinom{2n}{n}$ is an integer for all integers $n \geq 1$.

P2008-7 Find matrices B and C such that $B^3 + C^3 = \begin{bmatrix} 1 & -1 \\ 0 & 5 \end{bmatrix}$.

The 2008 Problem Solutions begin on page 99 in Chap. 31.

Chapter 9
2009 Problems

P2009-1 Suppose that the points P and Q are randomly selected in the interval $[0, 2]$, so the segment from P to Q has length $|\overline{PQ}| \geq 0$. What is the probability $\Pr[|\overline{PQ}| \leq \frac{1}{3}]$?

P2009-2 Show that the Maclaurin series of

$$f(x) = \frac{x}{1 - x - x^2}$$

is equal to $\sum_{n=1}^{\infty} f_n x^n$, where $f_1 = 1$, $f_2 = 1$, and $f_n = f_{n-1} + f_{n-2}$.

P2009-3 Let $T = \{1, 2, 3, 4, 5, 6, 7, 8\}$. Let the set S be defined as follows.

$$S = \{f | f : T \to T \text{ is a bijection}\}$$

with binary operation function composition. Let $\sigma \in S$. Suppose that σ^3 is defined as follows.

$$\sigma^3(1) = 2$$
$$\sigma^3(2) = 3$$
$$\sigma^3(3) = 5$$
$$\sigma^3(4) = 6$$
$$\sigma^3(5) = 7$$
$$\sigma^3(6) = 1$$

$$\sigma^3(7) = 8$$
$$\sigma^3(8) = 4$$

What is σ?

P2009-4 Show that $\sqrt[3]{2+\sqrt{5}} + \sqrt[3]{2-\sqrt{5}} = 1$.

P2009-5 Suppose that n is a composite number, $n > 0$, and $n \neq 4$. Show that $n | (n-1)!$.

P2009-6 Given $a, b \in \mathbb{R}$ with $a < b$, and a function $f : (a, b) \to \mathbb{R}$, suppose that f is increasing and satisfies the property that for all $\lambda \in (0, 1)$ and $x, y \in (a, b)$,

$$f(\lambda x + (1-\lambda)y) \leq \lambda f(x) + (1-\lambda)f(y).$$

Prove that f is continuous on (a, b).

P2009-7 Let $\mathbb{Z}_{\geq a}$ be equal to the set $\{x | x \in \mathbb{Z}, x \geq a\}$. It is known that there is a one-to-one correspondence

$$F : \mathbb{Z}_{\geq 0} \times \mathbb{Z}_{\geq 0} \times \mathbb{Z}_{\geq 0} \to \mathbb{Z}_{\geq 1}.$$

Write down a formula for F.

The 2009 Problem Solutions begin on page 103 in Chap. 32.

Chapter 10
2010 Problems

P2010-1 Suppose that you have a deck of n cards, divided into two stacks. One stack contains p red suit cards and the other stack contains q black suit cards, with $p + q = n$. You completely shuffle the stacks together. How many red–black color rearrangements of the cards are possible, assuming p and q are fixed?

P2010-2 Suppose that p is an odd prime and that $0 \leq k \leq p - 1$. Prove that

$$(p - (k+1))! k! \equiv (-1)^{k+1} \pmod{p}.$$

P2010-3 Consider the following game. There are two players, player 1 and player 2. There is a (non-empty) pile of coins, each identical, on the gaming table. Player 1 acts first and must remove either 1, 2, or 3 coins. Player 2 acts next and must remove either 1, 2, or 3 coins. The players continue taking turns in the manner described until there are no coins left on the gaming table. The player who selects the last coin is the loser. It is known that player 1 has a strategy that will guarantee a win if the number of chips on the table is equivalent to 0, 2, or 3 modulo 4. Explain what this strategy is, and prove that the strategy will guarantee the win for player 1.

P2010-4 Suppose that A and B are $n \times n$ matrices with real entries with the following two properties:

$$\operatorname{rank}(A) + \operatorname{rank}(B) = n$$

and

$$A + B = I,$$

where I is the $n \times n$ identity matrix. Prove that $A^2 = A$, $B^2 = B$, and $AB = BA$.

P2010-5 Evaluate the following limit.

$$\lim_{x \to 0} \frac{\sin(\arctan(x)) - \tan(\arcsin(x))}{\arcsin(\tan(x)) - \arctan(\sin(x))}.$$

P2010-6 Consider the lattice $\mathbb{Z} \times \mathbb{Z}$. Imagine that there is a point at the origin. The point will move in one of four directions in the lattice: up (the y coordinate is increased by one unit), right (the x coordinate is increased by one unit), left (the x coordinate is reduced by one unit), or down (the y coordinate is reduced by one unit). The probability that the point will move up is $1/4$, the probability that it will move down is $1/8$, the probability that it will move right is $1/16$, and the probability that the point will move left is $9/16$. Suppose that we keep track of the movement of the point for 5 rounds. What is the probability that it winds up in the open sector (not including the points along the line) in the first quadrant defined by the lines $y = 3x$ and $y = \frac{1}{2}x$?

P2010-7 A rational number can be written $\frac{p}{q}$ where p and q are integers, $q > 0$, and p and q have no common factors. Let the function $f(x)$ be defined as follows: If x is irrational, then $f(x) = 0$. If x is rational, then $f(x) = \frac{1}{q}$. Prove that this function is continuous at every irrational.

The 2010 Problem Solutions begin on page 109 in Chap. 33.

Chapter 11
2011 Problems

P2011-1 Suppose that children are playing a game together, where an even number, $n \geq 6$, of them are seated in a circle, facing the center. One of these children is designated as child number 1. The child to the right of child number 1 is designated as child number 2, and so on around the circle in a counterclockwise fashion. There is another child in the center of the circle, who is able to choose a positive integer i (which can be larger than n), known as the *elimination parameter*. Starting with child number 1, the child in the center of the circle counts off the first i children in a counterclockwise fashion (moving around the circle more than once if necessary). The child following the ith child leaves the circle and the game. The pattern continues, counting another i children starting with the next child in the circle after the child who has just left, until only one child remains in the circle, and that child is declared the winner. One can make a list of the children as they leave the game, establishing an order of elimination. Is there some n so that for any ordered list of $n - 1$ children from the circle, the child in the center can choose an elimination parameter i that realizes this list as an elimination order?

P2011-2 Suppose that we have a sheet of cardboard in the shape of an equilateral triangle of side length 1. You cut the corners off of the triangle by making a cut of length x units perpendicular to each side. At that point, the sides are then folded up and the box is formed. Show that the volume of the box is maximized when the area of the base of the box is positive and equal to the total area of the sides of the box.

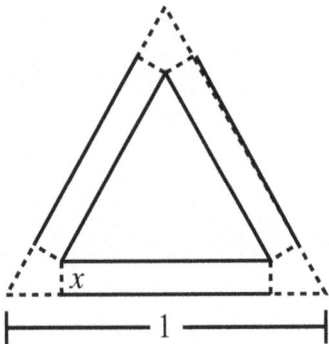

P2011-3 Let n be a positive integer. Consider, for every integer i with $1 \leq i \leq n$, the set C_i of real degree i polynomials with leading coefficient $\neq 0$ and constant term $= 0$, defined as follows:

$$C_i = \{g(x) = a_0 x^i + a_1 x^{i-1} + \cdots + a_{i-1} x \mid a_0 \neq 0, a_1, \ldots, a_{i-1} \in \mathbb{R}\}.$$

Let $C^n = C_1 \cup C_2 \cup \ldots \cup C_n$. Note that in particular for $i = 1$, $C^1 = C_1$ is this set of linear polynomials, $\{g(x) = a_0 x \mid a_0 \neq 0\}$.

Consider the following claim, depending on n: If $f : \mathbb{R}^2 \to \mathbb{R}$ is a function such that for every $g \in C^n$

$$\lim_{x \to 0} f(x, g(x)) = 0,$$

then

$$\lim_{(x,y) \to (0,0)} f(x, y) = 0.$$

Find some positive integer n for which the claim is true, and prove the claim, or show that the claim is false for all positive integers n by stating a specific counterexample $f(x, y)$ for each n.

P2011-4 Let b be a positive integer greater than 1. If x is a number, then there is an expansion of x with base b:

$$x = \sum_{i \in \mathbb{Z}} a_i b^i \quad 0 \leq a_i < b.$$

Let n be a positive integer such that $(n, b) = 1$. Show that the period of the expansion of $1/n$ is the smallest positive integer r such that $b^r \equiv 1 \mod n$.

P2011-5 Suppose that one has three pegs and two different stacks of n disks. Each stack of disks sits on a different peg (so there is one peg with no disks on it). The disks in one stack are painted red, and the disks in the other stack are painted blue. Each disk in the red stack has a different radius. The disk at the top of the red stack has the smallest radius, the second topmost disk has the second smallest radius, and so on all the way down to the bottom of the stack (which will be the disk with the largest radius). For each disk in the red stack, there is a disk in the blue stack identical to it except for color. The disks in the blue stack are arranged in exactly the same way as the disks in the red stack. The goal is to exchange the stacks of disks. There are some rules which limit possible moves. One can only move one disk at a time, and one cannot stack a larger disk upon a smaller disk. One is allowed to stack two disks which have the same size, but are different colors. Devise an algorithm which will exchange the position of the two stacks, and calculate the number of moves that your algorithm requires to perform the exchange.

P2011-6 Suppose that $f : \mathbb{R} \to \mathbb{R}$ is a differentiable function and that $|f'(x)| \leq 1/2$ for every x. A number x is a *fixed point* for $f(x)$ means that $f(x) = x$. Let x_1 be an arbitrary real number. Suppose that we define a sequence as follows. Let $x_2 = f(x_1)$; let $x_3 = f(x_2)$, and in general, let $x_n = f(x_{n-1})$. Show that the following limit exists:

$$\lim_{n \to \infty} x_n.$$

Also, show that the limit is a fixed point for the function $f(x)$.

P2011-7 Let A and B be two sets. A function

$$f : A \to B$$

is a *one-to-one correspondence* means that f is both one-to-one and onto. Show that there is a one-to-one correspondence between $\mathbb{R}$ and $\mathbb{R} \setminus \{0\}$.

P2011-8 For an $n \times n$ matrix A with complex entries, a vector $v \in \mathbb{C}^n$ is an *eigenvector* for A means that $v \neq \vec{0}$ and $Av = \lambda v$ for some $\lambda \in \mathbb{C}$. For $n \geq 2$, characterize the eigenvectors of the following $n \times n$ matrix:

$$A = \begin{bmatrix} 1 & 2 & 3 & \cdots & n \\ n & 1 & 2 & \cdots & n-1 \\ n-1 & n & 1 & \cdots & n-2 \\ \vdots & \vdots & \vdots & & \vdots \\ 2 & 3 & 4 & \cdots & 1 \end{bmatrix}.$$

The 2011 Problem Solutions begin on page 113 in Chap. 34.

Chapter 12
2012 Problems

P2012-1 Show that n^2 divides $(n+1)^n - 1$ for any positive integer n.

P2012-2 How many zeros are at the end of 213!?

P2012-3 Let $p(x) = a_n x^n + \cdots + a_2 x^2 + a_1 x + a_0$ be a polynomial of degree $n \geq 2$, with integer coefficients:

(a) Show that if a_0, a_1, a_n, and $a_2 + \cdots + a_n$ are odd integers, then $p(x)$ has no rational root.
(b) Give examples to show that the conclusion from part (a) may not be true if any of a_0, a_1, a_n, or $a_2 + \cdots + a_n$ is even.

P2012-4 Let A be an $n \times n$ matrix whose diagonal entries are all equal to the same real number $\alpha \in \mathbb{R}$ and all other entries are equal to $\beta \in \mathbb{R}$. Show that A is diagonalizable, and compute the determinant of A.

P2012-5 Let G be a group of order 26. Show that if G has a normal subgroup of order 2, then G is a cyclic group.

P2012-6 Show that for any positive integer k, the following is an irrational number:

$$\sum_{n=0}^{\infty} \frac{1}{(n!)^k}.$$

P2012-7 Let $f : [0, 1] \to \mathbb{R}$ be a continuous function. Show that

$$\lim_{n\to\infty} \int_0^1 nx(1-x^2)^n f(x)\,dx = \frac{1}{2} f(0).$$

P2012-8 Recall that a function $f(x, y)$ is *harmonic* in an open subset $\mathcal{O}$ of the plane means that it is twice continuously differentiable in $\mathcal{O}$ and $f_{xx}(x, y) + f_{yy}(x, y) = 0$ for all (x, y) in $\mathcal{O}$. Let $\mathcal{R}$ be the region in the plane given by

$$\mathcal{R} = \{(x, y) \in \mathbb{R}^2 : x^2 + (y+1)^2 \le 9 \text{ and } x^2 + (y-1)^2 \ge 1\}.$$

Show that if f is harmonic in an open disk $\mathcal{O}$ containing $\mathcal{R}$, then

$$\iint_{\mathcal{R}} f(x, y)\,dx\,dy = 9\pi f(0, -1) - \pi f(0, 1).$$

The 2012 Problem Solutions begin on page 121 in Chap. 35.

Chapter 13
2013 Problems

P2013-1 Define a sequence (s_n) recursively as follows: Let $s_1 = 1$, and for $n \geq 1$, let $s_{n+1} = \sqrt{1 + s_n}$. Prove that (s_n) converges, and then find the limit.

P2013-2 Let $\mathcal{C}$ be a non-empty collection (possibly infinite) of compact subsets of $\mathbb{R}$:

(a) Prove that $K = \bigcap_{C \in \mathcal{C}} C$ is a compact set.
(b) Give an example that illustrates that the union of a family of compact sets need not be compact.

P2013-3 Assume A and B are two sets with m and n elements, respectively:

(a) How many one-to-one functions are there from A to B?
(b) How many one-to-one and onto functions are there from A to B?

P2013-4 Let p and q be distinct prime numbers. Find the number of generators of the group $\mathbb{Z}_{pq}$.

P2013-5 Let G be a group and H a subgroup of G with index $(G : H) = 2$. Prove that H is a normal subgroup of G.

P2013-6 The Fibonacci numbers are defined as

$$f_1 = f_2 = 1$$

and

$$f_{n+1} = f_n + f_{n-1}$$

for $n \geq 2$:

(a) List $f_1, f_2, \ldots, f_7$.
(b) Illustrate, using the list from (a), that $f_{2n+1} = f_{n+1}^2 + f_n^2$ for $n = 1, 2, 3$.
(c) Prove that $f_{2n+1} = f_{n+1}^2 + f_n^2$ for all $n \in \mathbb{N}$.

P2013-7 Let a, b, m, M be real numbers with $0 < m \leq a \leq b \leq M$. Prove that

$$\frac{2\sqrt{mM}}{m+M} \leq \frac{2\sqrt{ab}}{a+b}.$$

P2013-8 A soccer ball is stitched together using white hexagons and black pentagons. Each pentagon borders five hexagons. Each hexagon borders three other hexagons and three pentagons. Each vertex is of valence 3 (meaning that at each corner of a hexagon or pentagon, exactly three hexagons or pentagons meet). How many hexagons and how many pentagons are needed to make a soccer ball? Hint: Euler's formula for convex polyhedra states that $V - E + F = 2$, where V is the number of vertices, E is the number of edges (i.e., the line adjoining two vertices), and F is the number of faces (hexagons or pentagons).

The 2013 Problem Solutions begin on page 131 in Chap. 36.

Chapter 14
2014 Problems

P2014-1 Let $a > 0$, and define the following function:

$$f(x) = \frac{\sqrt{a^3 x} - a\sqrt[3]{a^2 x}}{a - \sqrt[4]{ax^3}}. \qquad (14.1)$$

(a) Calculate these limits.

$$\lim_{x \to 0^+} f(x) =$$
$$\lim_{x \to a} f(x) =$$
$$\lim_{x \to +\infty} f(x) =$$

(b) Find the maximum value of $f(x)$ on its domain.

P2014-2 Let f be a function with domain $(0, \infty)$ satisfying:

- $f(x) = f(x^2)$ for all $x > 0$.
- $\lim_{x \to 0^+} f(x) = \lim_{x \to +\infty} f(x) = f(1)$.

Show that $f(x)$ is a constant function on $(0, \infty)$.

P2014-3 Let V be a corner of a right-angled box and let x, y, z be the angles formed by the long diagonal and the face diagonals starting at V. For

$$A = \begin{bmatrix} \sin x & \sin y & \sin z \\ \sin z & \sin x & \sin y \\ \sin y & \sin z & \sin x \end{bmatrix}$$

show that $|\det(A)| \leq 1$.

P2014-4 Let $f(t)$ be a real-valued integrable function on $[0, 1]$, so that both sides of the following equation are continuous functions of x:

$$2x - 1 = \int_0^x f(t)dt.$$

Prove that if $f(t) \leq 1$ for $0 \leq t \leq 1$, then there exists a unique solution $x \in [0, 1]$ of the equation.

P2014-5 Let $ABCD$ be a rectangle. The bisector of the angle ACB intersects AB at point M and divides the rectangle $ABCD$ into two regions: the triangle MBC with area s and the convex quadrilateral $MADC$ with area t:

(a) Determine the dimensions of the rectangle $ABCD$ in terms of s and t.
(b) In the case $t = 4s$, what is the ratio AB/BC?

P2014-6 In a badly overcrowded preschool, every child is either left-handed or right-handed, either blue-eyed or brown-eyed, and either a boy or a girl. Exactly half of the children are girls, exactly half of the children are left-handed, and exactly one fourth of the children are both. There are twenty-six children who are brown-eyed. Nine of those twenty-six are right-handed boys. Two children are right-handed boys with blue eyes. Thirteen children are both left-handed and brown-eyed. Five of these thirteen are girls:

(a) How many students does the preschool have?
(b) How many girls are right-handed and blue-eyed?

P2014-7 Let $n > 1$ be an integer. Let $(G, \cdot)$ be a group, with an identity element e and an element $a \in G$ with $a \neq e$ and $a^n = e$. Let $(H, *)$ be a group, let $f : G \to H$ be an arbitrary function, and then define $F : G \to H$ by

$$F(x) = f(x) * f(a \cdot x) * f(a^2 \cdot x) * \ldots * f(a^{n-1} \cdot x).$$

(a) Show that if $f(G)$ is a subset of some Abelian subgroup of H, then F is not a one-to-one function.

(b) Let $(H, *)$ be the symmetric group $(S_3, \circ)$ (the six-element group of permutations of three objects). Give an example of $(G, \cdot)$, n, and a as above, and a function $f : G \to H$, so that the expression F is a one-to-one function.

P2014-8 Determine whether the following sum of real cube roots is rational or irrational.

$$\sqrt[3]{6+\sqrt{\frac{847}{27}}} + \sqrt[3]{6-\sqrt{\frac{847}{27}}}.$$

The 2014 Problem Solutions begin on page 137 in Chap. 37.

Chapter 15
2015 Problems

P2015-1 An integer n has a *super-3 representation* means that there is a positive integer m, a sequence of distinct nonnegative integers $p_1, \ldots, p_m$, and a sequence $a_1, \ldots, a_m$ where each a_k is ± 1, so that

$$n = \sum_{k=1}^{m} a_k \cdot 3^{p_k} = a_1 \cdot 3^{p_1} + \cdots + a_m \cdot 3^{p_m}.$$

For instance, the integer 8 has the super-3 representation $8 = 3^2 - 3^0$, and the integer -11 has the super-3 representation $-11 = -3^2 - 3^1 + 3^0$. The number 0 has the *empty* super-3 representation, i.e., where $m = 0$ and the sum has no terms:

(a) Give a super-3 representation of 2015.
(b) Prove that every integer n has a super-3 representation.

P2015-2 Prove that, for every positive integer n,

$$\sum_{k=0}^{n} \sum_{i=0}^{n-k} \binom{n}{k} \binom{n-k}{i} 2^{k+i} = 5^n.$$

P2015-3 Three pairwise perpendicular line segments $\overline{AB}$, $\overline{CD}$, and $\overline{EF}$ have endpoints all on a sphere of unknown radius, and intersect inside the sphere at a point X. Given the lengths $AX = 1$, $CX = 2$, $EX = 3$, and $BX = 4$, determine, with proof, the volume of the octahedron with vertices A, B, C, D, E, and F.

P2015-4 You are standing in a room, which we will call Σ, which contains eight light switches numbered 1 through 8, all in the off position. On the other side of the door is another room, Λ, which contains eight lights, labeled A through H. Each switch controls exactly one light. Your goal is to determine which switches control which lights. You do this by making a number of trials, which consist of putting some set of switches in room Σ in the on position, and then entering room Λ to discover which lights are on. For instance, one trial might be to turn switches 1 and 3 to the on position (and all others off) and observe which lights in room Λ are on (perhaps lights D and H, though of course you cannot know this ahead of time):

(a) Give a strategy for finding out which switches control which lights using the smallest possible number of trials. (For this part, you do not need to prove that the number of trials you use is minimal.)
(b) Prove that your strategy uses the minimal number of trials; that is, prove that there is no strategy that determines which switches control which lights using fewer trials.

P2015-5 A function $f : \mathbb{R} \to \mathbb{R}$ is *smooth* means that all of its derivatives (first, second, third, etc.) exist everywhere:

(a) Suppose $f : \mathbb{R} \to \mathbb{R}$ is a smooth function with infinitely many zeros in the interval $[0, 1]$. Show that there is some $x \in [0, 1]$ such that $f^{(n)}(x) = 0$ for every integer $n \geq 0$ (that is, such that f and all its derivatives vanish at x).
(b) Give an example of such a function f which is nonconstant on every interval. (You need not prove that your function works.)

P2015-6 Recall that a graph is *planar* means that it can be drawn on the plane in such a way that no two of its edges cross. Further recall that a graph is *c-colorable* (for a positive integer c) means that its vertices can be colored with c colors in such a way that no two adjacent vertices have the same color. The famous *Four Color Theorem* says that every planar graph can be 4-colored.

A graph is *k-color-planar* means that it can be drawn on the plane, and its *edges* colored with k colors, in such a way that no two edges with the same color cross. Thus, a 1-color-planar graph is just a planar graph.

Shown below are two graphs illustrating these definitions. Graph G is planar and its vertices can be 3-colored as shown, with the three colors represented by •, ○, and ⊗. The graph H is 2-color-planar (but not planar), with the two colors represented by thin and thick edges.

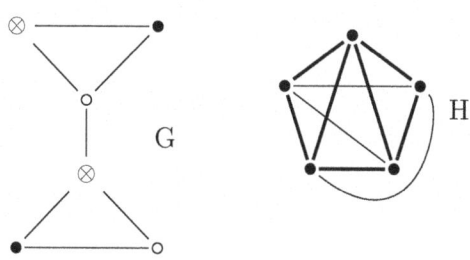

Prove that, for every positive integer k, there is a positive integer c_k such that every k-color-planar graph is c_k-colorable. Hint: You may use the Four Color Theorem in your proof if you wish.

P2015-7 Let S be a non-empty, finite set and $* : S \to S$ be a binary operation on S. Suppose that $*$ satisfies the following two conditions:

- $*$ is associative; that is, for any $a, b, c \in S$, $(a * b) * c = a * (b * c)$.
- For any $a, b \in S$, $a * (a * (b * a)) = b$:

(a) Prove that $*$ is commutative; that is, that for any $a, b \in S$, $a * b = b * a$.
(b) Prove that $|S|$, the size of S, is a power of 3.

P2015-8 Let $f : \mathbb{R} \to \mathbb{R}$ be twice differentiable on $[0, 1]$, and suppose that $f(0) = f'(0) = 0$ and $f(1) = 1$. Prove that there is some $a \in (0, 1)$ such that $f'(a) f''(a) = \dfrac{9}{8}$.

The 2015 Problem Solutions begin on page 145 in Chap. 38.

Chapter 16
2016 Problems

P2016-1 Assume we are looking at numbers in normal decimal representations, and consider the set

$$A = \{1, 11, 111, 1111, \ldots\} = \{x \in \mathbb{Z} \mid x \text{ consists entirely of 1s}\}.$$

For all $a \in A$, define $n(a)$ to be the number of 1s in a's representation. (For example, $n(1) = 1$ and $n(111) = 3$.) Prove or disprove the following statement: The set $\{a \in A \mid n(a) \text{ divides } a\}$ is infinite.

P2016-2 Let $f : \mathbb{Z} \to \mathbb{Z}$ be a function with all of the following properties:

(i) $f(2) = 2$.
(ii) $f(mn) = f(m)f(n)$ for all $m, n \in \mathbb{Z}$.
(iii) If $m > n$, then $f(m) > f(n)$.

Conjecture what all the possible candidates for f are, and then prove your conjecture.

P2016-3 Arrange eight points in 3-space so that each of the 56 possible triplets of points determined forms an isosceles triangle. Prove your arrangement works. (Note: Some triangles may be degenerate!)

P2016-4 Let A be a subset of $\mathbb{R}$, and let $f, g : A \to A$ be two continuous functions. The function f is *homotopic* to g means that there is a continuous function $h : A \times [0, 1] \to A$ such that $h(a, 0) = f(a)$ and $h(a, 1) = g(a)$ for all a in A. (This function is sometimes called a deformation function. It may be helpful to think of $[0, 1]$ as being a "slider" that continuously morphs the function f into the function g.)

(a) Let $A = [0, 1] \subset \mathbb{R}$, and define $f, g : A \to A$ by $f(a) = a^2$ and $g(a) = a^3$. Find a function h demonstrating that f is homotopic to g.
(b) Recall that an equivalence relation $\sim$ on a set X requires three properties:

- *Reflexivity*: $\forall x \in X, x \sim x$.
- *Symmetry*: $\forall x, y \in X$, if $x \sim y$, then $y \sim x$.
- *Transitivity*: $\forall x, y, z \in X$, if $x \sim y$ and $y \sim z$, then $x \sim z$.

Prove that for any subset A of $\mathbb{R}$, being homotopic forms an equivalence relation on the set $C(A)$ of all continuous functions $f : A \to A$.

P2016-5 The Fibonacci numbers f_n ($n = 0, 1, 2, \ldots$) are defined recursively by $f_0 = 0$, $f_1 = 1$, and $f_n = f_{n-2} + f_{n-1}$ for $n \geq 2$:

(a) Find a linear transformation $T : \mathbb{R}^2 \to \mathbb{R}^2$ such that $T(f_n, f_{n+1}) = (f_{n+1}, f_{n+2})$ for all $n \geq 0$, and then state and prove a conjecture for what $T^n(0, 1)$ equals for all $n \geq 1$.
(b) Find the matrix A of T with respect to the standard basis for $\mathbb{R}^2$, and then find the eigenvalues of A.
(c) Find a non-recursive expression for f_n for all $n \geq 1$.

P2016-6 Let A be an open subset of the real numbers and $f : A \to \mathbb{R}$ be a differentiable function such that $f'(a) = 0$ for all $a \in A$:

(a) Prove that if $A = \mathbb{R}$, then f must be a constant function. (Note: Do not use antiderivatives in your proof, as the proof that antiderivatives are unique up to a constant depends on this statement.)
(b) Prove or disprove: If $A \neq \mathbb{R}$, then f must be a constant function.

P2016-7 Let G be a set and $* : G \times G \to G$ be a binary operation. G, together with the operation $*$, is a *group* means that the following conditions are satisfied:

- *Associativity*: $\forall a, b, c \in G, (a * b) * c = a * (b * c)$.
- *Identity*: $\exists e \in G$ such that $\forall g \in G, e * g = g * e = g$.
- *Inverse*: $\forall a \in G, \exists a^{-1} \in G$ such that $a * a^{-1} = a^{-1} * a = e$.

For purposes of shorthand, the symbol $*$ is often omitted (i.e., we write $a * b$ as simply ab and $a * a * a$ as simply a^3).

Assume that G is a group which has the following additional properties:

(i) For all $g, h \in G, (gh)^3 = g^3 h^3$.
(ii) There are no elements in G of order 3 (i.e., $\nexists g \in G$ such that $g^3 = e$).

16 2016 Problems

For any group G with properties (i) and (ii) listed above:

(a) Prove that for all $a, b \in G$, if $a \neq b$, then $a^3 \neq b^3$.
(b) Prove that the map $\phi : G \to G$ defined by $\phi(g) = g^3$ is a bijection if G is finite.

P2016-8 *Remark from the problem authors*: The length of this puzzle problem is for clarity, not necessarily its difficulty.

Assume elections are conducted by voters who place all the candidates in rank order. Under these conditions there are several possible voting methods available:

- Plurality—the candidate with the most first place votes wins
- Plurality with elimination—the following process is repeated until there is a winner:
  - If there is a candidate that has more than half the first place votes, he/she wins.
  - Otherwise, eliminate the candidate(s) with the fewest first place votes. Reorder the voters' preferences by moving other candidates up in the ranks to fill the vacancies.
- Borda count—Candidates receive points based upon each voter's ranking. A last-place vote earns a candidate one point, a second-to-last-place vote earns a candidate two points, a third-to-last-place vote earns a candidate three points, and so forth, with a first place vote earning a candidate n points, where n is the total number of candidates. The candidate with the most total points wins.
- Pairwise Comparison—Candidates receive points based upon their performance in a round-robin style analysis. Each candidate is paired with each other candidate. (For example, if there were four candidates, there would be six possible matchups.) For each matchup, count the number of voters who prefer one candidate over the other. The winner of the matchup (more voter preferences) receives a point; if there is a tie, both candidates receive half a point. The candidate with the most total points wins.
- Survivor—The following process is repeated until there is a winner:
  - If there is only one candidate remaining, he/she wins.
  - Otherwise, eliminate the candidate(s) with the most last-place votes. Reorder the voters' preferences by moving other candidates up in the ranks to fill the vacancies.

Example Consider the following sample preference table for nine voters in an election with three candidates: X, Y, and Z:

Rank/Number of Votes	4	3	2
First	X	Y	Z
Second	Y	Z	Y
Third	Z	X	X

That is, four voters rank X as their first choice, Y as their second choice, and Z as their third choice; three voters would pick Y first, Z second and X third; and two voters would choose Z first, Y second, and X third. Using this table of preferences, the winners for the various voting methods would be:

- Plurality—X wins with the most first place votes (4 of them).
- Plurality with elimination—No candidate has more than half the first place votes, so we eliminate the candidate(s) with the fewest. Candidate Z is eliminated. The new preference table is

Rank/Number of Votes	4	3	2
First	X	Y	Y
Second	Y	X	X

  Now Y wins with more than half the first place votes.
- Borda count—The point totals are as follows:

  - X: $4(3)+3(1)+2(1)=17$
  - Y: $4(2)+3(3)+2(2)=21$
  - Z: $4(1)+3(2)+2(3)=16$

  Therefore, Y wins.
- Pairwise comparison—The matchups are as follows:

  - X vs. Y: Y wins 5 to 4 because 5 voters like Y better than X; Y gets a point.
  - X vs. Z: Z wins 5 to 4 because 5 voters like Z better than X; Z gets a point.
  - Y vs. Z: Y wins 7 to 2 because 7 voters like Y better than Z; Y gets a point.

  Therefore, Y wins.
- Survivor—We eliminate the candidate(s) with the most last-place votes. Candidate X is eliminated. The new preference table is

Rank/Number of Votes	4	3	2
First	Y	Z	Y
Second	Z	Y	Z

  Now Z is eliminated, leaving only Y remaining; therefore, Y wins.

Challenge: Construct a preference table for an election with five candidates—A, B, C, D, and E—such that A wins via plurality, B wins via plurality with elimination,

C wins via Borda count, D wins via pairwise comparison, and E wins via survivor. You may use any number of voters you would like.

The 2016 Problem Solutions begin on page 151 in Chap. 39.

Chapter 17
2017 Problems

P2017-1 In constructible geometry, one constructs points, lines, and circles from given points, lines, and circles, using an unmarked straightedge and compass:

- The straightedge draws a line between points already given, which includes the line segment connecting them; the line may extend as far beyond either point as desired. New points are created where the line intersects already existing lines or circles.
- The compass draws a circle (or a circular arc) centered on a given point with a radius extending to another point from the center. Again, new points are created where the circle intersects other circles or lines:

(a) Use a straightedge and compass to construct the midpoint M of the line segment $\overline{AB}$ given below. (Show your construction and label M. Do not erase any intermediate steps.)

(b) Use a straightedge and compass to construct a square $ABCD$ with the line segment $\overline{AB}$ given below as one of its sides. (Show your construction and labels C and D. Do not erase any intermediate steps.)

(c) Given an isosceles right triangle ABC, one can use a compass to construct lunes as follows. First, one semicircle is formed with the line segment $\overline{AC}$ as its diameter. Two other semicircles are formed with $\overline{AB}$ and $\overline{BC}$ as their diameters. The lunes are the shaded shapes in the figure below.

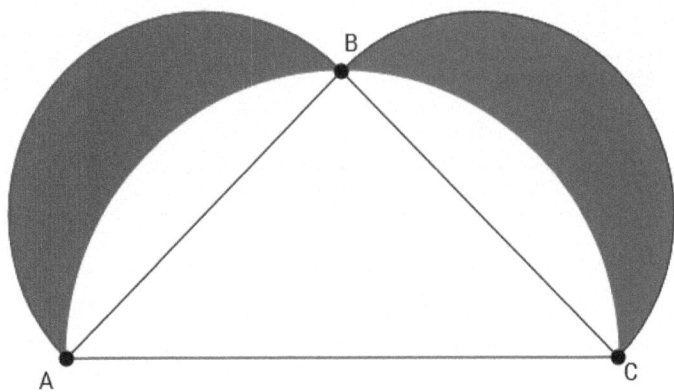

Use a straightedge and compass to construct a square whose area is equal to the area of **one** of the lunes on the diagram given. Justify how you know the square you construct has the appropriate area.

P2017-2 Consider the following series

$$\sum_{n=1}^{\infty} (a_n)^n,$$

where

$$a_n = \begin{cases} \frac{1}{3^1} + \frac{1}{3^2} + \cdots + \frac{1}{3^n} & \text{if } n \text{ is odd,} \\ |\sin(n)\cos(n)| & \text{if } n \text{ is even.} \end{cases}$$

Make a conjecture as to whether the series converges or not, and then prove your conjecture.

P2017-3 Two students, Joe and Frank, are each asked to independently select a number at random from the interval [0,1] using a uniform probability distribution (where the probability of selecting a number from a subinterval $[a, b]$ is equal to the length $b - a$). If a denotes the number chosen by Joe and b denotes the number chosen by Frank, what is the probability that the quadratic equation $x^2 + ax + b = 0$ has at least one real root?

P2017-4 A *multiplicative magic square* is an $n \times n$ square array of numbers consisting of n^2 distinct positive integers (not necessarily consecutive) arranged such that the product of the n numbers in any of the n rows, n columns, or two main diagonal lines is always the same number. Call this common product the *magic product*:

(a) Show that the magic product of a 3×3 multiplicative magic square must be a perfect cube.
(b) Find an example of a 3×3 multiplicative magic square whose magic product is minimal. Explain how you know this magic product is minimal.

P2017-5 For any positive integer n, let $s(n)$ be the sum of the first n terms of the sequence

$$0, 1, 1, 2, 2, 3, 3, 4, 4, \ldots, k, k, k+1, k+1, \ldots$$

(a) Find a formula for $s(n)$. (Note: Your final formula should not have "..." in it.)
(b) Suppose that m and n are any two positive integers with $m > n$. Prove that $s(m+n) - s(m-n) = mn$.

P2017-6 Suppose that A is an $n \times n$ matrix such that every entry of A is ± 1. Show that the determinant of A is divisible by 2^{n-1}.

P2017-7 A robot is programmed to shuffle cards in such a way so that it always rearranges cards in the same way relative to the order in which the cards are given to it. The thirteen hearts arranged in the order

$$A, 2, 3, 4, 5, 6, 7, 8, 9, 10, J, Q, K$$

are given to the robot, shuffled, and then the shuffled cards are given back to the robot and shuffled again. This process is repeated until the cards have been shuffled a total of 7 times. At this point the order of the cards is

$$2, 4, 6, 8, 10, Q, A, K, J, 9, 7, 5, 3.$$

What was the order of the cards after the first shuffle?

P2017-8 Recall that a subset A of $\mathbb{R}$ is *closed* in $\mathbb{R}$ means that the complement $\mathbb{R} \setminus A$ is equal to a union of any number of open intervals (possibly infinitely many). Suppose A is a non-empty, closed subset of $\mathbb{R}$ such that for each $a \in A$, every open interval that contains a also contains another element of A. Show that A must be uncountable.

The 2017 Problem Solutions begin on page 159 in Chap. 40.

Chapter 18
2018 Problems

P2018-1 Show that
$$\sin(x)\sin(2x)\cdots\sin(nx) \neq 1$$
for every real number x and any positive integer $n \geq 2$.

P2018-2 Determine the smallest natural number n such that
$$\frac{1}{1+\sqrt{2}} + \frac{1}{\sqrt{2}+\sqrt{3}} + \cdots + \frac{1}{\sqrt{n}+\sqrt{n+1}} \geq 100.$$

P2018-3 Suppose $a_1, a_2, \ldots, a_n$ are strictly positive real numbers and
$$a_1^x + a_2^x + \cdots + a_n^x \geq n,$$
for every real number x. Prove that $a_1 a_2 \cdots a_n = 1$.

P2018-4 Show that if $x + y + z > 0$, then
$$\det \begin{bmatrix} x & z & y \\ y & x & z \\ z & y & x \end{bmatrix} \geq 0.$$

P2018-5 Consider the following sequence defined recursively:
$$x_1 = \frac{1}{2}, \quad x_{k+1} = x_k^2 + x_k, \quad k \geq 1.$$

Find the integer part of S_{100}, where:

$$S_{100} = \frac{1}{x_1 + 1} + \frac{1}{x_2 + 1} + \cdots + \frac{1}{x_{100} + 1}.$$

P2018-6 Consider a semicircle and AB its diameter. Pick two arbitrary points D and E on the semicircle such that the segments AD and BE intersect at M in the interior of the semicircle. Prove that:

$$|AM| \cdot |AD| + |BM| \cdot |BE| = |AB|^2.$$

P2018-7 Suppose a fair six-sided die is rolled, and let X denote the outcome of the die roll. Suppose a fair coin is flipped until X heads are observed. Compute the expected value of the number of flips.

P2018-8 A town has n inhabitants who like to form clubs. They want to form clubs so that every pair of clubs should share a member, but no three clubs should share a member. What is the maximum number of clubs they can form? Illustrate with an example.

The 2018 Problem Solutions begin on page 167 in Chap. 41.

Chapter 19
2019 Problems

P2019-1 Provide counterexamples to each of the following statements:

(a) If both $f(x)$ and $g(x)$ are continuous and monotone on $\mathbb{R}$, then $f(x) + g(x)$ is continuous and monotone on $\mathbb{R}$.
(b) If a function $f(x)$ is not bounded in any neighborhood of a point a, then either $\lim_{x \to a^+} |f(x)| = \infty$ or $\lim_{x \to a^-} |f(x)| = \infty$.
(c) A function cannot be continuous at only one point in its domain and discontinuous everywhere else.
(d) If a function is differentiable and increasing on an interval (a, b), then its derivative is positive on the interval (a, b).
(e) If $f(x)$ is a function with an antiderivative $F(x)$ that is defined at both a and b, then $\int_a^b f(x)dx = F(b) - F(a)$.

P2019-2 Let $f(x) = x^2/4$, and consider the set of all right triangles in the plane whose right angle vertex lies at the origin and whose other two vertices lie somewhere else on the graph of $y = f(x)$. Conjecture on the existence and location of a point other than the origin through which all such triangles must pass, and then prove your conjecture.

P2019-3 Two-Face and his henchmen are once again on a crime spree terrorizing the good people of Gotham City. However, rather than flipping a coin to decide his behavior, Two-Face has decided to adopt a more deterministic approach. When Two-Face's gang crashes a party with 60 guests in attendance, they first line everyone up against the wall. Then, the first henchman walks down the line and takes $10 from every guest. Next, a second henchman walks down the line giving $10 to the second, fourth, sixth, etc., party guests. This process is repeated with a third henchman taking $10 from the third guest and every third guest thereafter, then with a fourth henchman giving $10 to the fourth guest and every fourth guest thereafter, and so on, until finally the sixtieth henchman gives $10 to only the sixtieth (last) person in line:

(a) How many party guests made money? How many lost money? How many broke even?
(b) Which party guest(s) made the most money?
(c) Which party guest(s) lost the most money?
(d) How much money did Two-Face and his henchmen make?

P2019-4 A *fair* coin is a coin that will produce a result of either heads (H) or tails (T) when flipped with equal probability. Suppose that you start flipping a fair coin and record the sequence of H/T results:

(a) What is the expected number of flips needed to get your first tail (T)?
(b) What is the expected number of flips needed to achieve your first string of heads followed by tails (HT)?
(c) What is the expected number of flips needed to achieve your first string of heads followed by heads (HH)?

P2019-5 Let $A = \begin{bmatrix} 1 & 2019 \\ 0 & 1 \end{bmatrix}$. State a conjecture on the four entries of the matrix A^n for any $n \in \mathbb{Z}$, and then prove your conjecture.

P2019-6 Let $(s_n)_{n=1}^{\infty}$, $(t_n)_{n=1}^{\infty}$, and $(u_n)_{n=1}^{\infty}$ be sequences with the following properties:

- (s_n) is monotone decreasing.
- (t_n) is monotone increasing.
- $s_n \geq u_n \geq t_n$ for every $n \in \mathbb{N}$.

For each of the three sequences $(s_n)_{n=1}^{\infty}$, $(t_n)_{n=1}^{\infty}$, and $(u_n)_{n=1}^{\infty}$, conjecture that the sequence must converge must not converge, or not enough information is given to determine convergence. Then prove your conjectures.

P2019-7 Suppose that G is a group. A subset S of G is a set of *generators* for G means that every element of G can be written as a finite product of elements in S and/or their inverses. Suppose that G is a group with identity e which has $\{x, y\}$ as a set of generators, where the generators x and y satisfy the following relations:

- $x^2 = e$.
- $y^4 = e$.
- $xyxyxy = e$.

Determine the maximum number of elements that G can contain. Justify your answer.

P2019-8 Suppose that T is the triangular pyramid with vertices at $(0, 0, 0)$, $(12, 0, 0)$, $(0, 8, 0)$, and $(0, 0, 24)$. What is the maximum volume that a rectangular prism R which has one vertex at $(0, 0, 0)$ and which is inscribed in T can have?

The 2019 Problem Solutions begin on page 173 in Chap. 42.

Chapter 20
2020 Problems

The year 2020 had enough problems.

In the spring of 2020, the Indiana Section of the MAA was scheduled to hold its annual spring meeting in conjunction with an interdisciplinary, multi-organization conference hosted by IUPUI to celebrate the 150th anniversary of the founding of Indiana University in Bloomington. Unfortunately, the COVID-19 pandemic erupted just a few months before the conference was scheduled, and the State of Indiana declared a public health emergency and a stay-at-home order. The larger conference was canceled, so the Section meeting was canceled, and consequently, the 2020 edition of the Indiana College Mathematics Competition was canceled.

Chapter 21
2021 Problems

P2021-1 Let $A = \begin{bmatrix} 1 & 2021 \\ 0 & 1 \end{bmatrix}$.

(a) Find A^{2021}.
(b) Find a 2×2 matrix B so that $B^{2021} = A$.

P2021-2 The polynomial $p(x) = 16x^4 - 32x^3 - 104x^2 + 122x + 232$ has an interesting property. There is a line $y = mx + b$ that is tangent to the graph of p in *two* places. Find the line. Hint: Consider the polynomial $q(x) = p(x) - (mx + b)$ and how it might factor.

P2021-3 A three-digit positive integer n is exactly five times the product of its digits.

(a) Show that the digits of n must all be odd.
(b) Find n.

P2021-4 Consider the points $A(3, 4, 1)$, $B(5, 2, 9)$, and $C(1, 6, 5)$ in $\mathbb{R}^3$. Show that these points are the vertices of a cube.

P2021-5 Let $S = \{1, 2, 3, 4, 5, 6, 7, 8\}$. A partition P of S into two-element subsets $\{\{x_1, y_1\}, \{x_2, y_2\}, \{x_3, y_3\}, \{x_4, y_4\}\}$ has each x_i and each y_i a value from S, with all values used precisely once.

(a) How many partitions of S into four two-element subsets are there?

(b) For a given partition, define its *value* by $V(P) = \sum_{i=1}^{4} x_i y_i$. An integer n is *achievable* means that there is a partition P whose value $V(P)$ is n. Find the minimum and maximum achievable values.

P2021-6 Let $\underline{xyz}$ be a three-digit number, made with digits x, y, and z. That is, $\underline{xyz} = 100x + 10y + z$.

(a) Find digits a, b, and c, not necessarily distinct, for which $\underline{abc} + \underline{cab} - \underline{bca} = \underline{608}$.
(b) Show that there are no values of a, b, and c that satisfy $\underline{abc} + \underline{cab} - \underline{bca} = \underline{707}$.

P2021-7 Polynomial p has nonnegative integer coefficients and satisfies $p(1) = 21$ and $p(11) = 2021$. What is $p(10)$?

P2021-8 In a "KenKen" puzzle, the numbers in each heavily outlined set of squares, called cages, must combine (in any order) to produce the target number in the top corner of the cage using the mathematical operation indicated. A number can be repeated within a cage as long as it is not in the same row or column. In the 5 × 5 puzzle, each of the digits 1 through 5 must appear precisely once in each row and column. Solve the 5 × 5 KenKen below.

The 2021 Problem Solutions begin on page 181 in Chap. 43.

Chapter 22
2022 Problems

P2022-1 Consider the four lines $l_1 : x = -3$, $l_2 : x = 1$, $l_3 : y = 2$, and $l_4 : y = -4$. If A is some point in the plane, suppose each of the segments from A to the lines meets perpendicularly at B, C, D, and E, respectively (i.e., B is on l_1, C on l_2, etc.). Consider the locus of all points A where

$$|AB||AC| = |AD||AE|.$$

Find an equation describing this locus, and specify as much as possible the type of plane curve it is.

P2022-2 Find two nonzero functions $f(x)$ and $g(x)$ so that $f'(x) \neq 0$, $g'(x) \neq 0$, and

$$\frac{d}{dx}(f(x)g(x)) = f'(x)g'(x).$$

P2022-3 Consider the function $f(x) = \dfrac{1}{1 - e^{-1/x}}$.

(a) Assuming $x > 0$, find $f'(x)$.

(b) Compute $\displaystyle\int_0^1 \frac{e^{-1/x}}{x^2(1 - e^{-1/x})^2}\, dx$.

P2022-4 Let A be a square matrix, and suppose positive integers m and n exist so that $A^m = I$ and $A^n \neq I$. Find

$$\det(I + A + A^2 + \ldots + A^{m-1}).$$

P2022-5

(a) Define $n?$ as the sum of integers from 1 to n, for example, $5? = 1+2+3+4+5$. Compute the number of zeros that appear at the end of decimal representation of $2022?$.
(b) Define $n!$ as the product of integers from 1 to n, for example, $5! = 1*2*3*4*5$. Compute the number of zeros that appear at the end of the decimal representation of $2022!$.

P2022-6 Can a group be the union of two of its proper subgroups?

P2022-7 For a function $g : [0, 1] \to [0, 1]$, denote $g \circ g \circ ... \circ g$ (m times) as g^m. Suppose that $g : [0, 1] \to [0, 1]$ is continuous and that there is an m so that for all x, $g^m(x) = x$. Show that in fact $g^2(x) = x$.

P2022-8 It is well known that $\mathbb{N} = \{1, 2, 3, ...\}$ and $\mathbb{N} \times \mathbb{N}$ have the same cardinality, and the standard classroom demonstration of this involves a diagonal lines argument. Explicitly give a function between $\mathbb{N}$ and $\mathbb{N} \times \mathbb{N}$, and show that it is bijective.

The 2022 Problem Solutions begin on page 187 in Chap. 44.

Chapter 23
2023 Problems

P2023-1 A particle is moving along a track. At time t (in seconds) its distance from one end of the track is $x(t) = 9 + 7t + 3t^2 - t^3/3$ (in feet). What is the fastest speed that the particle goes during the time interval $0 \le t \le 10$, and when is this? What is the slowest speed the particle ever goes, and when is this?

P2023-2 Verify that the following equation is true:

$$2\cos 70° = \frac{\sin 50°}{\sin 60°} + \frac{\sin 60°}{\sin 50°} - \frac{\sin^2 70°}{\sin 50° \sin 60°}.$$

P2023-3 For real numbers x, y, and z, define $f(x, y, z)$ to be the maximum of x, y, and z, that is, $f(x, y, z) = \max\{x, y, z\}$. Evaluate

$$\int_0^1 \int_0^1 \int_0^1 f(x, y, z)\, dx\, dy\, dz.$$

P2023-4 Find a differentiable function f such that

$$\int_{-x}^{x} f(t)\, dt = f(x) - f(-x)$$

and $f(0) = 2023$.

P2023-5 Let $f_1, f_2, f_3, \ldots$ be the Fibonacci sequence $1, 1, 2, 3, 5, \ldots$, in which $f_{n+1} = f_n + f_{n-1}$. Prove that

$$f_{n+1} f_{n-1} - f_n^2 = (-1)^n \qquad \text{for all } n \ge 2.$$

P2023-6 Consider $\triangle ABC$ with its inscribed circle γ. Assume that the altitude $\overline{CH}$ does not pass through the center of γ and that H is between A and B. Let X and Y be the intersections of $\overline{CH}$ and γ as shown. Prove that $CX > HY$.

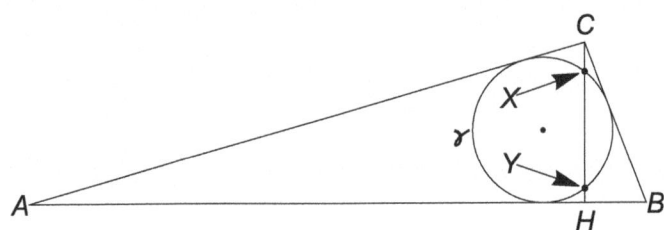

P2023-7 You have four playing cards. You verify that three of them are red suits and one is black. You give them to your friend. While you are not watching, your friend mixes them randomly and places them face down in four locations on a table. You choose one of the cards. You tell your friend your choice, but you do not look at the card. Your friend looks at the three cards you did not choose and turns one of the red cards face up. For each card still face down, what is the probability that it is red?

P2023-8 There are n small white beads and k black beads moving along an infinite wire without gravity or friction. The white beads are all to the left of the black beads. (The figure shows the case $n = 3, k = 5$.) Initially the white beads move to the right, and the black beads move to the left, all with the same speed. When two beads collide, they bounce, reversing directions but keeping the same speed.

(a) After the collisions are done, how many beads will be moving to the right and how many beads will be moving to the left?
(b) How many collisions will there be?

Note: This is not meant to be a physics problem, so please avoid concepts such as mass, energy, and momentum in your answer.

The 2023 Problem Solutions begin on page 193 in Chap. 45.

Part II
Solutions

Chapter 24
2001 Solutions

The 36th competition was held at the University of Indianapolis.

The 2001 problem set was prepared by Professor Mike Axtell at Wabash College and Professor Joe Stickles at the University of Evansville.

The **2001 problem statements** begin on page 3 in Chap. 1.

S2001-1 The sum of the first a positive integers is given by the sequence of triangular numbers,

$$(0, 1, 3, 6, 10, \ldots, T_a, \ldots),$$

defined by $T_a = \frac{1}{2}a(a+1)$ or recursively by $T_0 = 0$ and $T_{a+1} = T_a + a + 1$. Then the sum of any list of consecutive positive integers is a difference of triangular numbers: For $0 \leq a < b$,

$$(a+1) + (a+2) + (a+3) + \cdots + (b-1) + b \tag{24.1}$$
$$= T_b - T_a$$
$$= \frac{1}{2}b(b+1) - \frac{1}{2}a(a+1)$$
$$= \frac{1}{2}(b+a+1)(b-a). \tag{24.2}$$

For a positive integer n which is not a power of 2, there are some numbers $k \geq 0$ and $m \geq 1$ so that $n = 2^k(2m+1)$. The factors 2^k and $2m+1$ can correspond to the two factors from (24.2), so solving for a and b leads to these two cases:

Case 1. If $m+1 \leq 2^k$, then the sum of the positive numbers from $2^k - m$ to $2^k + m$ is

$$T_{2^k+m} - T_{2^k-m-1}$$
$$= \frac{1}{2}((2^k+m)+(2^k-m-1)+1)((2^k+m)-(2^k-m-1))$$
$$= \frac{1}{2}(2^{k+1})(2m+1) = n.$$

Case 2. If $m+1 > 2^k$, then the sum of the positive numbers from $m+1-2^k$ to $m+2^k$ is

$$T_{2^k+m} - T_{m-2^k}$$
$$= \frac{1}{2}((2^k+m)+(m-2^k)+1)((2^k+m)-(m-2^k))$$
$$= \frac{1}{2}(2m+1)(2^{k+1}) = n.$$

In either case, there are at least two consecutive numbers in the sum (24.1); in particular, if n is odd, then $k = 0$ and $n = m + (m+1)$. In (24.2), one of the factors must be odd, so if the product is power of 2, then $b - a = 1$ and the sum (24.1) has only one term.

Remark from the Editors This well-known problem appears in [RL], where it is attributed to a 1976 Canadian Olympiad.

S2001-2 Inflection points of smooth functions are where $f''(x)$ changes sign. For $x > 0$, $x^{1/x}/x^4 > 0$, so we must determine the number of times that $g(x) = ((\ln(x))^2 + (2x-2)\ln(x) + (1-3x))$ changes sign. Let $h(x) = g(e^x) = x^2 + (2e^x - 2)x + 1 - 3e^x$ for $x \in \mathbb{R}$. $h(x)$ changes sign when $g(e^x)$ changes sign. We see that $h'(x) = (e^x+1)(2x-1) - 1$ and $h''(x) = e^x(2x+1) + 2 > 0$, so h' is increasing and has at most one zero. $h'(1/2) = -1 < 0$ and $h'(1) = e > 0$, so $h'(x)$ has exactly one zero. By Rolle's Theorem, $h(x)$ has at most two zeros. $h(-1) = 4 - \frac{5}{e} > 0$, $h(0) = -2 < 0$, and $h(2) = 1 + e^2 > 0$. Therefore $h(x)$ changes sign exactly twice. Hence $f''(x)$ changes sign exactly twice, and $f(x)$ has exactly two inflection points.

S2001-3 Drop one ball into each slot, leaving $k - n$ balls. Now the question is how many ways can the remaining $k - n$ balls be allocated into the n slots. This problem is in one-to-one correspondence with arrangements of the $k - n$ balls along with $n - 1$ "dividers."

To see an example of this, presume $k = 8$ and $n = 4$, meaning there are four balls remaining, which we will represent with an o, and three "dividers," which we will represent with an x. Note that the arrangement

OOXOXOX

would correspond with adding two balls to the first slot, one to the second, one to the third, and zero to the fourth slot. Similarly, the arrangement

XXOOOOX

would correspond with adding zero balls to the first slot, zero to the second, four to the third, and zero to the fourth slot. In this case, the number of possible arrangements would be $\binom{7}{3}$.

In general, the number of arrangements is the number of ways one can select the $n-1$ dividers from the $(k-n)+(n-1) = k-1$ places, which is $\binom{k-1}{n-1}$.

S2001-4A

(a) We apply Euler's Identity, $e^{i\theta} = \cos(\theta) + i\sin(\theta)$ to θ and 3θ to get

$$\cos(3\theta) + i\sin(3\theta) = e^{3i\theta} = (e^{i\theta})^3$$
$$= (\cos(\theta) + i\sin(\theta))^3$$
$$= \cos^3(\theta) - 3\cos(\theta)\sin^2(\theta)$$
$$+ i\left(3\cos^2(\theta)\sin(\theta) - \sin^3(\theta)\right).$$

Therefore

$$\cos(3\theta) = \cos^3(\theta) - 3\cos(\theta)\sin^2(\theta) \qquad (24.3)$$
$$= \cos^3(\theta) - 3\cos(\theta)(1 - \cos^2(\theta))$$
$$= 4\cos^3(\theta) - 3\cos(\theta).$$

Thus $f(x) = 4x^3 - 3x$.

(b) $\frac{1}{2} = \cos(\pi/3) = f(\cos(\pi/9))$. Thus $\cos(\pi/9)$ is a root of the polynomial with integer coefficients $p(x) = 2(4x^3 - 3x - 1/2)$, which can be rewritten as a composite $g(2x) = (2x)^3 - 3(2x) - 1$. By the Rational Root Theorem, any rational root of $g(x) = x^3 - 3x - 1$ must be ± 1. Since $g(1) = -3$ and $g(-1) = 1$, $g(2x)$ has no rational roots, and $\cos(\pi/9)$ is irrational.

S2001-4B This alternate approach to part (a) does not use complex numbers but assumes knowledge of the angle addition formula for cosine and the double angle formula for sine.

$$\cos(3\theta) = \cos(2\theta + \theta)$$
$$= \cos(2\theta)\cos(\theta) - \sin(2\theta)\sin(\theta)$$
$$= (\cos^2(\theta) - \sin^2(\theta))\cos(\theta) - (2\sin(\theta)\cos(\theta))\sin(\theta)$$
$$= \cos^3(\theta) - 3\sin^2(\theta)\cos(\theta),$$

which matches (24.3).

Remark from the Editors The statement of the Rational Root Theorem is that if a degree n polynomial with integer coefficients $a_n x^n + \ldots + a_1 x + a_0$ has a rational root equal to a fraction in the lowest terms r/s, then r is a divisor of a_0, and s is a divisor of a_n. A discussion of this theorem appears in the solution to Problem #1 of the 1907 Eötvös Competition in Hungary [K2] and also in [RL].

S2001-5 Let $Q = (0, 0)$ be the center of the larger circle C_2, and without loss of generality we may rotate the circle so that the path passes through $(2, 0)$. Let O_t be the point at the center of C_1 at time t, with position vector $\overrightarrow{QO_t}$ parametrized by $(\cos(t), \sin(t))$ for $0 \le t \le 2\pi$. At time $t = 0$, set $P = (2, 0)$, so $\angle QO_0P = \pi$ radians. In one revolution around Q, the smaller circle rolls along a path twice its circumference, so at time t, $\angle QO_tP = \pi - 2t$ as in the figure, and $\overrightarrow{O_tP}$ is in the direction $(\cos(-t), \sin(-t))$. Adding vectors,

$$\overrightarrow{QP} = \overrightarrow{QO_t} + \overrightarrow{O_tP} = (\cos(t), \sin(t)) + (\cos(-t), \sin(-t)) = (2\cos(t), 0),$$

so the point P oscillates on a diameter segment.

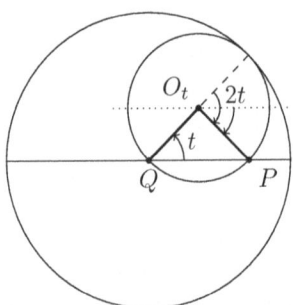

Remark from the Editors This problem is similar to Problem #3 of the 1926 Eötvös Competition in Hungary [K2].

S2001-6 We observe that for each integer k the interval $[k\pi + \pi/3, k\pi + 2\pi/3]$ has length $\pi/3 > 1$ and so must contain at least one integer. The absolute value of that integer is less than $4k$ when $k > 1$. For the integers inside $[k\pi + \pi/3, k\pi + 2\pi/3]$

for some k, we see that $\dfrac{|\sin(n)|}{n} > \dfrac{\sqrt{3}/2}{4k}$. Hence

$$\sum_{n=1}^{\infty} \dfrac{|\sin(n)|}{n} \geq \sum_{k=2}^{\infty} \dfrac{\sqrt{3}/8}{k}.$$

Since the Harmonic Series diverges to infinity, the comparison test shows that $\sum_{n=1}^{\infty} \dfrac{|\sin(n)|}{n}$ is infinite.

Chapter 25
2002 Solutions

The 37th competition was held at Anderson University.

The 2002 problem set was prepared by Professor Mike Axtell at Wabash College and Professor Joe Stickles at the University of Evansville.

The **2002 problem statements** begin on page 5 in Chap. 2.

S2002-1

(a) An example of such a function would be $f(x) = x - \frac{b}{2}$.
(b) Note that the assumption that f is positive and continuous on $[0, b]$ implies that the expression is well defined and integrable.

$$\int_0^b \frac{f(x)}{f(x) + f(b-x)} dx = \int_0^b \frac{f(x) + f(b-x) - f(b-x)}{f(x) + f(b-x)} dx$$

$$= \int_0^b \left(1 - \frac{f(b-x)}{f(x) + f(b-x)}\right) dx$$

$$= b - \int_0^b \frac{f(b-x)}{f(x) + f(b-x)} dx.$$

Now, let $u = b - x$. Then, $du = -dx$. Also, when $x = 0$, we have $u = b$, and when $x = b$, we have $u = 0$. Hence, we have

$$\int_0^b \frac{f(x)}{f(x) + f(b-x)} dx = b - \int_b^0 \frac{f(u)}{f(b-u) + f(u)} (-du)$$

$$\Longrightarrow \int_0^b \frac{f(x)}{f(x) + f(b-x)} dx = b - \int_0^b \frac{f(u)}{f(u) + f(b-u)} du$$

$$\implies \int_0^b \frac{f(x)}{f(x)+f(b-x)}\,dx + \int_0^b \frac{f(u)}{f(u)+f(b-u)}\,du = b.$$

The variables make no difference in the integrals, so the left-hand side becomes

$$2\int_0^b \frac{f(x)}{f(x)+f(b-x)}\,dx = b$$

$$\implies \int_0^b \frac{f(x)}{f(x)+f(b-x)}\,dx = \frac{b}{2}.$$

Remark from the Editors This problem and solution are similar to Problem #5 in [HWi].

S2002-2 Since each $a_i > 0$ for all $i \geq 1$, we have

$$b_n = \frac{a_1 + a_2 + \cdots + a_n}{n} \geq \frac{a_1}{n} = a_1 \cdot \frac{1}{n}.$$

Since the series $\sum \frac{1}{n}$ diverges, the series $\sum \frac{a_1}{n}$ diverges (using $a_1 \neq 0$), and hence $\sum b_n$ diverges by the direct comparison test.

S2002-3 Let p and q be consecutive odd primes with $p < q$. Then, $p+q$ is even and can be factored into these two positive integers:

$$p+q = 2\left(\frac{p+q}{2}\right).$$

Now, the fraction $\frac{p+q}{2}$ represents the average of p and q, and since $p < q$, we have $p < \frac{p+q}{2} < q$. Since these two primes were consecutive, we must have $\frac{p+q}{2}$ is composite, and hence it has at least two (possibly repeated) prime factors. Thus, the sum of two consecutive odd primes has at least three prime factors. These factors may repeat, considering the examples $3+5 = 2^3$, $5+7 = 2^2 \cdot 3$, and $7+11 = 2 \cdot 3^2$, or there may be exactly three prime factors, $13+17 = 2 \cdot 3 \cdot 5$.

Remark from the Editors This problem appears, with a similar solution, as Problem #80 in [HWi].

S2002-4 Let a_n denote the total number of books sold from day 1 through day n. Hence, we have

$$1 \leq a_1 < a_2 < a_3 < \cdots < a_{100} = 140.$$

Adding 59 to each above term gives

$$60 \leq a_1 + 59 < a_2 + 59 < a_3 + 59 < \cdots < a_{100} + 59 = 199.$$

Now, the list $a_1, a_2, \ldots, a_{100}, a_1 + 59, a_2 + 59, \ldots, a_{100} + 59$ is a list of 200 numbers that lie between 1 and 199, inclusive. By the pigeonhole principle (also known as Dirichlet's box principle), there must be two numbers in the list that have the exact same value. Since $a_1 < a_2 < \cdots < a_{100}$, we know $a_i \neq a_j$ for any $i \neq j$. Similarly, we cannot have $a_i + 59 = a_j + 59$ for $i \neq j$. Hence, we must have $a_i = a_j + 59$ for some i, j, with $i > j$ by the increasing property of the list, and $a_i - a_j = 59$. Because $a_i - a_j$ represents the number of books sold from day $j+1$ to day i, there were exactly 59 books sold from day $j+1$ until day i.

Remark from the Editors This problem and solution are similar to Problem #22 in [HWi].

S2002-5 In some steps, multiplication is denoted with the · (dot).

(a) Let e be the identity of S. If there exists an $x \in S$ such that $\varphi(x) = 0$, then

$$\varphi(e) = \varphi\left(xx^{-1}\right) = \varphi(x)\varphi\left(x^{-1}\right) = 0 \cdot \varphi\left(x^{-1}\right) = 0,$$

and for any other $y \in S$, we have

$$\varphi(y) = \varphi(ey) = \varphi(e)\varphi(y) = 0 \cdot \varphi(y) = 0,$$

and hence $\varphi(x) = 0$ for all $x \in S$.
Now, suppose $\varphi(x) \neq 0$ for all $x \in S$. Then,

$$\varphi(e) = \varphi(ee) = \varphi(e)\varphi(e)$$
$$\Longrightarrow 0 = \varphi(e)\varphi(e) - \varphi(e) = (\varphi(e) - 1)\varphi(e)$$
$$\Longrightarrow \varphi(e) = 1 \text{ (since } \varphi(e) \neq 0\text{),}$$

and for any $x \in S$, we have

$$\varphi(e) = \varphi\left(xx^{-1}\right)$$
$$\Longrightarrow 1 = \varphi(x)\varphi\left(x^{-1}\right).$$

Since $\varphi(x)$ and $\varphi\left(x^{-1}\right)$ are nonnegative integers with product 1, we must have $\varphi(x) = 1$. Hence, $\varphi(x) = 1$ for all $x \in S$.

(b) If $\varphi(0_S) \neq 0$, then for all $x \in S$, we get

$$\varphi(0_S) = \varphi(0_S \cdot x) = \varphi(0_S)\varphi(x)$$

$$\Longrightarrow \varphi(x) = 1,$$

which contradicts φ being nonconstant. Hence, we must have $\varphi(0_S) = 0$.

S2002-6 First, for the four corners of an inscribed square to lie on the three sides of the triangle, two corners must lie on the same side, and they cannot be opposite corners, so (at least) one square side is a subinterval of one triangle side as in the figure. Next, we need the fact that for any numbers b and h, we have

$$(b-h)^2 \geq 0$$
$$b^2 - 2bh + h^2 \geq 0$$
$$b^2 + h^2 \geq 2bh.$$

Now, we notice that the square "divides" the triangle into a trapezoid with bases b and s and a height of s and a triangle with base s and height $h-s$. We must have the sum of these areas equal the total area of the triangle; hence, we have

$$\frac{1}{2}s(b+s) + \frac{1}{2}s(h-s) = \frac{1}{2}bh$$
$$bs + s^2 + sh - s^2 = bh$$
$$s(b+h) = bh$$
$$s^2(b+h)^2 = b^2h^2$$
$$s^2\left(b^2 + 2bh + h^2\right) = b^2h^2$$
$$s^2(2bh + 2bh) \leq b^2h^2 \text{ by above}$$
$$s^2(4bh) \leq b^2h^2$$
$$s^2 \leq \frac{1}{4}bh = \frac{1}{2}\left(\frac{1}{2}bh\right).$$

Hence, the area of the square is at most one half the area of the triangle.

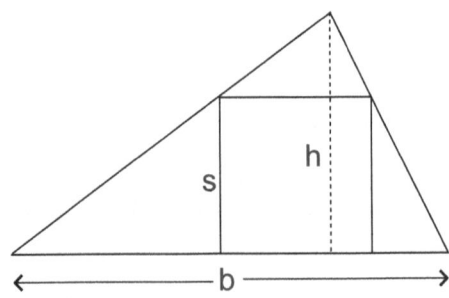

Chapter 26
2003 Solutions

The 38th competition was held at Butler University.

The 2003 problem set was prepared by Professor Mike Axtell at Wabash College and Professor Joe Stickles at the University of Evansville.

The **2003 problem statements** begin on page 7 in Chap. 3.

The 2003 competition's first place team from Indiana—Purdue Fort Wayne. Left to Right: Jeff Wilkins, Christian MacLeod, Coach Doug Weakley, and Kevin Chlebik. Photo courtesy of Purdue Fort Wayne Communications and Marketing [PFW]

S2003-1

(a) Since $(a * b) * (a * b) = a * b$ by the given property, we have by substitution that $c * c = c$.

(b) We have $(a * b) * (c * y) = a * y$ by the given property. Using the hypothesis, we get $c * (c * y) = a * y$. From part (a), we can substitute $c * c$ in for c, yielding $(c * c) * (c * y) = a * y$. Finally, using the given property, we get $c * y = a * y$.

S2003-2 Let p_k represent the probability a particular pair of competitors will face each other in a tournament involving 2^k competitors. By induction on k, $k = 1$: There are only 2^1 competitors, so $p = 1 = \frac{1}{2^{k-1}}$. Assume $p_k = \frac{1}{2^{k-1}}$ for $k \leq n$. For $k = n+1$, the probability a competitor meets any other competitor in the first round is $\frac{1}{2^{k+1}-1}$. The probability they meet in later rounds is the probability they do not meet in round 1, they both win in round 1, and they meet in subsequent rounds. This is given by $\left(1 - \frac{1}{2^{k+1}-1}\right)\left(\frac{1}{2}\right)\left(\frac{1}{2}\right) p_n$. So,

$$p_{n+1} = \frac{1}{2^{n+1}-1} + \left(1 - \frac{1}{2^{k+1}-1}\right)\left(\frac{1}{2}\right)\left(\frac{1}{2}\right) p_n$$

$$= \frac{1}{2^{n+1}-1} + \frac{2^{n+1}-2}{2^{n+1}-1} \cdot \frac{1}{4} \cdot \frac{1}{2^{n-1}} \text{ by ind. hyp.}$$

$$= \frac{1}{2^{n+1}-1} + \frac{2^n - 1}{2^n(2^{n+1}-1)}$$

$$= \frac{2^n + 2^n - 1}{2^n(2^{n+1}-1)}$$

$$= \frac{1}{2^n}.$$

By induction, $p_k = \frac{1}{2^{k-1}}$ for all $k \geq 1$.

S2003-3A If we connect the five points on the unit circle, we get a regular pentagon. Hence, the length a is the same as the one shown here.

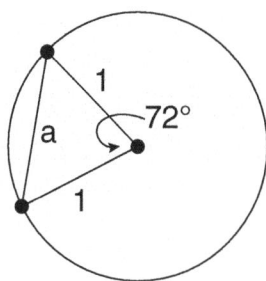

From the Law of Cosines, $a = \sqrt{2 - 2\cos 72°}$. Going back to the given chords, we get the following picture:

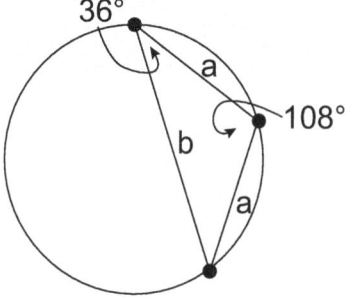

Again using the Law of Cosines, we get $b^2 = 2a^2 - 2a^2 \cos 108° = 4 - 4\cos 72° - (4 - 4\cos 72°)\cos 108° = 4 - 4\cos 72° + (4 - 4\cos 72°)\cos 72° = 4 - 4\cos^2 72° = 4\sin^2 72°$ so $b = 2\sin 72°$. Using the Law of Sines, we get

$$\frac{\sin 108°}{b} = \frac{\sin 36°}{a}$$

$$\frac{\sin 72°}{2\sin 72°} = \frac{\sin 36°}{a}$$

$$a = 2\sin 36°.$$

So,

$$ab = (2\sin 36°)(2\sin 72°) = 4(2\sin 18° \cos 18°)(\cos 18°)$$
$$= 8\sin 18° \cos^2 18° \qquad (26.1)$$
$$= 8\sin 18°(1 - \sin^2 18°).$$

S2003-3B In the first figure above, by drawing the diameter of the circle that bisects the segment of length a, one can find a right triangle with an $18°$ angle opposite to side length $\frac{1}{2}a$ and with hypotenuse b, so $\sin 18° = \frac{a}{2b}$. In the second figure, drawing the radius that bisects the segment of length b results in a right triangle with $18°$ angle adjacent to side length $\frac{1}{2}b$ and with hypotenuse 1, so $\cos 18° = \frac{b}{2}$. Then $ab = (2b \sin 18°)(2 \cos 18°)$, leading to the above line (26.1).

S2003-4 Write the square as an array:

a	b	c
d	e	f
g	h	i

Then $k = aei = beh = ceg$. So, $a = \dfrac{k}{ei}, b = \dfrac{k}{eh}, c = \dfrac{k}{eg}$. Now,

$$k = abc = \frac{k}{ei} \cdot \frac{k}{eh} \cdot \frac{k}{eig} = \frac{k^3}{e^3 ghi} = \frac{k^2}{e^3},$$

and solving for k gives $k = e^3$.

Remark from the Editors See **S2017-4** in Chap. 40 for an alternate solution.

S2003-5 Considering some partial sums,

$$S_3 = 1 + \left(\frac{1}{2} - \frac{1}{3}\right) \geq 1$$

$$S_6 = 1 + \left(\frac{1}{2} - \frac{1}{3}\right) + \frac{1}{4} + \left(\frac{1}{5} - \frac{1}{6}\right) \geq 1 + \frac{1}{4}$$

$$S_9 = 1 + \left(\frac{1}{2} - \frac{1}{3}\right) + \frac{1}{4} + \left(\frac{1}{5} - \frac{1}{6}\right) + \frac{1}{7} + \left(\frac{1}{8} - \frac{1}{9}\right) \geq 1 + \frac{1}{4} + \frac{1}{7}.$$

The pattern is $S_{3n} \geq 1 + \dfrac{1}{4} + \dfrac{1}{7} + \cdots + \dfrac{1}{3n-2}$. A limit comparison test with $\sum \dfrac{1}{n}$ shows $\sum \dfrac{1}{3n-2}$ diverges. Hence, a direct comparison shows the given series has unbounded partial sums and diverges.

S2003-6

(a) By induction on n, for $n = 2$, $Q^2 = \begin{bmatrix} 2 & 1 \\ 1 & 1 \end{bmatrix} = \begin{bmatrix} f_3 & f_2 \\ f_2 & f_1 \end{bmatrix}$. Assume $Q^n = \begin{bmatrix} f_{n+1} & f_n \\ f_n & f_{n-1} \end{bmatrix}$. For $n+1$, $Q^{n+1} = Q^n Q = \begin{bmatrix} f_{n+1} & f_n \\ f_n & f_{n-1} \end{bmatrix} \begin{bmatrix} 1 & 1 \\ 1 & 0 \end{bmatrix} = \begin{bmatrix} f_{n+2} & f_{n+1} \\ f_{n+1} & f_n \end{bmatrix}$.
So, the result holds.

(b) From part (a), $Q^{3n} = \begin{bmatrix} f_{3n+1} & f_{3n} \\ f_{3n} & f_{3n-1} \end{bmatrix}$. Also,

$$Q^{3n} = Q^n Q^n Q^n = \begin{bmatrix} f_{n+1} & f_n \\ f_n & f_{n-1} \end{bmatrix}^3.$$

Direct computation shows that the upper right entry when cubing this matrix is $(f_{n+1}^2 + f_n^2) f_n + f_{n+1} f_n f_{n-1} + f_n f_{n-1}^2$. Equating this with the upper right entry of the first matrix, we get

$$\begin{aligned}
f_{3n} &= (f_{n+1}^2 + f_n^2) f_n + f_{n+1} f_n f_{n-1} + f_n f_{n-1}^2 \\
&= f_{n+1}^2 f_n + f_n^3 + f_{n+1} f_n f_{n-1} + f_n f_{n-1}^2 \\
&= f_{n+1}^2 (f_{n+1} - f_{n-1}) + f_n^3 \\
&\quad + f_{n+1}(f_{n+1} - f_{n-1}) f_{n-1} + (f_{n+1} - f_{n-1}) f_{n-1}^2 \\
&= f_{n+1}^3 - f_{n+1}^2 f_{n-1} + f_n^3 + f_{n+1}^2 f_{n-1} - f_{n+1} f_{n-1}^2 + f_{n+1} f_{n-1}^2 - f_{n-1}^3 \\
&= f_{n+1}^3 + f_n^3 - f_{n-1}^3.
\end{aligned}$$

Chapter 27
2004 Solutions

The 39th competition was held at Indiana State University.

The 2004 problem set was prepared by Professor Mike Axtell at Wabash College and Professor Joe Stickles at the University of Evansville.

The **2004 problem statements** begin on page 9 in Chap. 4.

S2004-1 We attempt to partition $\{1, 2, 3, 4, 5\}$ into two sets A and B in such a way that neither set contains two numbers and their difference. Thus, 2 cannot be in the same set as either 1 or 4; else we would have $2 - 1 = 1$ or $4 - 2 = 2$. So, put 2 in A, and put 1 and 4 in B. If we put 3 in B, then we have $4 - 3 = 1$. So, 3 must go in A. Similarly, placing 5 in B leads to $5 - 4 = 1$; thus, 5 cannot be in B. However, 5 cannot be in A since $5 - 3 = 2$. We have reached a contradiction. Hence, no matter how the two sets are constructed, one of the two sets must contain two numbers and their difference.

Remark from the Editors This problem and solution appear as Problem #3 of the 1916 Eötvös Competition in Hungary [K_2].

S2004-2

(a) Since

$$\lim_{n\to\infty} \left| \left(\frac{2^{n+1}}{a^{2^{n+1}}+1}\right) / \left(\frac{2^n}{a^{2^n}+1}\right) \right| = \lim_{n\to\infty} \left(\frac{2^{n+1}}{2^n} \cdot \frac{a^{2^n}+1}{a^{2^{n+1}}+1}\right)$$

$$= 2 \lim_{n\to\infty} \left[(1 + \frac{1}{a^{2^n}})/(a^{2^n} + \frac{1}{a^{2^n}}) \right]$$

$$= 2(0) = 0,$$

this series converges by the Ratio Test.
(b) Since

$$\frac{2^n}{a^{2^n}+1} = \frac{2^n(a^{2^n}-1)}{(a^{2^n}+1)(a^{2^n}-1)}$$
$$= \frac{2^n(a^{2^n}-1) + 2^{n+1} - 2^{n+1}}{(a^{2^n}+1)(a^{2^n}-1)}$$
$$= \frac{2^n(a^{2^n}+1) - 2^{n+1}}{(a^{2^n}+1)(a^{2^n}-1)}$$
$$= \frac{2^n}{a^{2^n}-1} - \frac{2^{n+1}}{a^{2^{n+1}}-1},$$

the sum $\sum_{n=0}^{\infty} \frac{2^n}{a^{2^n}+1} = \sum_{n=0}^{\infty} \left(\frac{2^n}{a^{2^n}-1} - \frac{2^{n+1}}{a^{2^{n+1}}-1} \right)$ is telescoping (with the last terms in partial sums approaching 0 by part (a)). Hence

$$\sum_{n=0}^{\infty} \frac{2^n}{a^{2^n}+1} = \frac{2^0}{a^{2^0}-1} = \frac{1}{a-1}.$$

S2004-3 Let B be the matrix obtained from A by subtracting row 1 of A from each of the other three rows. Then $\det A = \det B$. Each entry in the last three rows of B is -3, 0, or 3 and therefore divisible by 3. Now let C be the matrix obtained from B by dividing each of the entries in the last three rows of B by 3. All of the entries of C are integers, giving $\det C$ is an integer; moreover, $\det A = \det B = 3^3 \det C$. So, $\det A$ is divisible by 27.

Remark from the Editors This problem and solution appear as Problem #10 in [GKL].

S2004-4 Induction on n, the length of the sequence, starts with $n = 1$, where the sequence is

$$a_1 = \frac{\sum_{i=0}^{1-1} \binom{1-1}{i} a_{i+1}}{2^{1-1}}.$$

So, assume that for any sequence of length k, with $k \leq n$, we will have final term

$$\frac{\sum_{i=0}^{k-1}\binom{k-1}{i}a_{i+1}}{2^{k-1}}.$$

For $k = n+1$, after the first step, we have the sequence

$$\frac{a_1+a_2}{2}, \frac{a_2+a_3}{2}, \ldots, \frac{a_n+a_{n+1}}{2},$$

which is a sequence of length n. Thus, by the inductive hypothesis, the final term will be

$$\frac{\sum_{i=0}^{n-1}\binom{n-1}{i}\left(\frac{a_{i+1}+a_{i+2}}{2}\right)}{2^{n-1}}$$

$$= \frac{\left(\sum_{i=0}^{n-1}\binom{n-1}{i}a_{i+1}\right) + \left(\sum_{i=0}^{n-1}\binom{n-1}{i}a_{i+2}\right)}{2^n}$$

$$= \frac{a_1 + a_{n+1} + \left(\sum_{i=1}^{n-1}\binom{n-1}{i}a_{i+1}\right) + \left(\sum_{i=0}^{n-2}\binom{n-1}{i}a_{i+2}\right)}{2^n}$$

$$= \frac{a_1 + a_{n+1} + \left(\sum_{i=0}^{n-2}\binom{n-1}{i+1}a_{i+2}\right) + \left(\sum_{i=0}^{n-2}\binom{n-1}{i}a_{i+2}\right)}{2^n}$$

$$= \frac{a_1 + a_{n+1} + \sum_{i=0}^{n-2}\left(\binom{n-1}{i+1} + \binom{n-1}{i}\right)a_{i+2}}{2^n}$$

$$= \frac{a_1 + a_{n+1} + \sum_{i=0}^{n-2}\binom{n}{i+1}a_{i+2}}{2^n}$$

$$= \frac{a_1 + a_{n+1} + \sum_{i=1}^{n-1}\binom{n}{i}a_{i+1}}{2^n}$$

$$= \frac{\sum_{i=0}^{n}\binom{n}{i}a_{i+1}}{2^n}.$$

S2004-5 Since P is the center of the square, $\triangle APC$ will also be a right triangle. Construct a circle with diameter $\overline{AC}$; both points B and P will be on this circle. (This is because the circle that circumscribes a right triangle has as its center the midpoint of the hypotenuse.) Since $\overline{AP}$ and $\overline{PC}$ are equal chords of this circle, the arcs AP and PC are equal. Thus $\angle ABP = \angle CBP$.

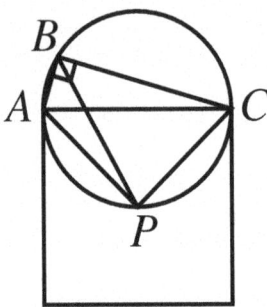

S2004-6 Since both boats remain in their slips for the same amount of time, this information does not enter into the solution of the problem. When the ferryboats meet for the first time, the combined distance the boats have traveled is equal to the width of the river. When the boats reach the opposite shore, the combined distance the boats have traveled equals two widths of the river. When they then meet the second time, the combined distances the boats have traveled are three widths of the river. Since the boats move at a constant speed, it follows that each boat has traveled three times as far as when they first met and had traveled a combined distance of one river width. The slow boat had traveled 720 yards when the boats first met. Thus, by the second meeting, the slow boat has traveled $3 \times 720 = 2160$ yards. Since this second meeting occurs at the point when the slow boat has moved 400 yards away from the far shore, it follows that the width of the river is given by $2160 - 400 = 1760$ yards.

Chapter 28
2005 Solutions

The 40th competition was held at Indiana University—Purdue University Fort Wayne for the first time.

The 2005 problem set was prepared by Professor Dan Coroian at the host institution.

The **2005 problem statements** begin on page 11 in Chap. 5.

Professor David Housman, Goshen College, presents a book prize to the 2005 first place team from Purdue University

S2005-1 Since x is in $[-1, 1]$, we can make the substitution $\theta = \cos^{-1} x$, with $\theta \in [0, \pi]$. The integral becomes

$$\int_{-1}^{1} \frac{|\sin(n \cos^{-1} x)|}{\sqrt{1-x^2}}\, dx = -\int_{\pi}^{0} |\sin n\theta|\, d\theta = \int_{0}^{\pi} |\sin n\theta|\, d\theta.$$

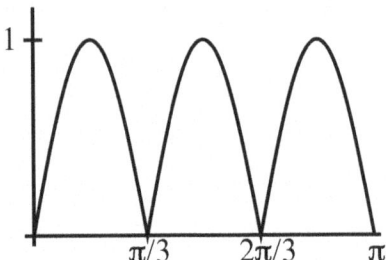

From the graph of $|\sin n\theta|$ (seen above for $n = 3$), the integral can be calculated as

$$\int_0^\pi |\sin n\theta|\, d\theta = n \int_0^{\pi/n} \sin n\theta\, d\theta = -n \frac{\cos n\theta}{n}\bigg|_0^{\pi/n} = 2.$$

S2005-2 Toward a contradiction, assume that there exists a red–white–blue coloring of the points in the plane such that no two points of the same color are at distance 1 from each other. We will show that for such a coloring there are no two points of different colors at a distance of $\sqrt{3}$ units. Indeed, suppose such a pair does exist. Without loss of generality, we can assume it is a red–white pair and denote the two points R and W, respectively. If we construct a rhombus $RBWA$ with side lengths 1 and diagonal RW, we see that the other two vertices, A and B, must be colored blue. However, it is easy to see that diagonal AB has length 1, which contradicts our original assumption. Therefore, for the coloring in question, all points at a distance $\sqrt{3}$ have the same color. This implies that the circle $x^2 + y^2 = 3$ will have to be colored the same color as the origin, which leads to a contradiction with our assumption that no two points of the same color are at distance 1 from each other (just choose two points on this circle that are distance 1 from each other).

S2005-3 We have

$$0 \le a_n = \frac{1}{n^3} \sum_{k=1}^n \ln(1+kn) = \frac{1}{n^3} \ln \prod_{k=1}^n (1+kn)$$

$$\le \frac{1}{n^3} \ln \prod_{k=1}^n (1+n^2) = \frac{1}{n^3} \ln(1+n^2)^n = \frac{1}{n^2} \ln(1+n^2)$$

$$\le \frac{1}{n^2} \ln(1+n)^2 = \frac{2}{n^2} \ln(1+n) \le \frac{2}{n^2}(1+n) = \frac{2n+2}{n^2}.$$

Thus, $0 \le a_n \le \dfrac{2n+2}{n^2}$, so using the Squeeze Theorem, $\lim\limits_{n \to \infty} a_n = 0$.

S2005-4A If we let $b = a$ and $d = 1/a$ in the first equation, we obtain $(a*a)(c*(1/a)) = (ac)*1 = ac$, that is, $c*(1/a) = ca$. If we now use $1/a$ instead of a, we obtain $c*a = c/a$, so operation $*$ is just the usual division of rational numbers. Therefore, $((6/5)*(8/15))*2 = ((6/5) \div (8/15)) \div 2 = (6/5)(15/8)(1/2) = 9/8$.

S2005-4B Lemma: $x*y = x \div y$. Proof: Step 1. $y \cdot (1*y) = (y*1) \cdot (1*y) = (y \cdot 1)*(1 \cdot y) = y*y = 1$, so $1*y$ is the reciprocal of y (by uniqueness of inverse in $\mathbb{Q}^+$). Step 2. $x*y = (x \cdot 1)*(1 \cdot y) = (x*1) \cdot (1*y) = x \cdot y^{-1}$. ∎

From the lemma,

$$((6/5)*(8/15))*2 = ((6/5) \div (8/15)) \div 2 = (6/5)(15/8)(1/2) = 9/8.$$

Remark from the Editors In both of the above solutions, showing that the operation is the same as $\div$ also shows the existence of such an operation $*$, so that the three given properties are not logically inconsistent.

S2005-5A Since $B = I - A$, $AB + I = A(I - A) + I = I + A - A^2$, and $I - A + A^2$ is an inverse:

$$\begin{aligned}(I + A - A^2)&(I - A + A^2) \\ &= (I - A + A^2) + A(I - A + A^2) - A^2(I - A + A^2) \\ &= I - A + A^2 + A - A^2 + A^3 - A^2 + A^3 - A^4 \\ &= I - A^2 + A^3 + A^3 - A^4.\end{aligned}$$

The last four terms cancel because $A^4 = A^3 A = A^2 A = A^3 = A^2$. One should also cite the linear algebra fact that in the ring $M_n(\mathbb{R})$ it is enough to check just one product or explicitly check that the product in the other order also gives the identity by an analogous computation (then, this solution works for any ring).

S2005-5B Multiply $A + B = I$ by A on each side, and use the assumption $A^2 = A^3$, to see that $ABA = 0$. Then $I - AB$ is an inverse:

$$(AB + I)(I - AB) = AB - (AB)(AB) + I^2 - AB = I - (ABA)B = I$$

(and similarly for the product in the other order).

S2005-6 We have

$$\sqrt{9x^4 - 24x^3 + 6x^2 + 5} - (ax^p + bx + c)$$

$$= x^2 \left[\sqrt{9 - \frac{24}{x} + \frac{6}{x^2} + \frac{5}{x^4}} - x^{p-2}\left(a + \frac{b}{x^{p-1}} + \frac{c}{x^p}\right) \right].$$

If $p \neq 2$ or if $p = 2$ and $a \neq 3$, then the limit is $\pm\infty$, so $p = 2$ and $a = 3$. Substituting these values and rationalizing the numerator the limit become

$$\lim_{x \to \infty} \frac{9x^4 - 24x^3 + 6x^2 + 5 - (3x^2 + bx + c)^2}{\sqrt{9x^4 - 24x^3 + 6x^2 + 5} + (3x^2 + bx + c)}$$

$$= \lim_{x \to \infty} \frac{(-24 - 6b)x^3 + (6 - b^2 - 6c)x^2 + \cdots}{x^2 \left(\sqrt{9 - \frac{24}{x} + \frac{6}{x^2} + \frac{5}{x^4}} + 3 + \frac{b}{x} + \frac{c}{x^2} \right)}.$$

For the limit to be equal to $7/3$, we must have $-24 - 6b = 0$ and $\frac{6-b^2-6c}{6} = \frac{7}{3}$, which gives $b = -4$ and $c = -4$.

S2005-7 Extend AM until it crosses DC in D', and extend BN until it crosses AD in A'. Triangle ABP is similar to $D'NP$ (since $D'N$ is parallel to AB). Then

$$\frac{BP}{NP} = \frac{AB}{ND'} = \frac{AB}{NC + CD'} = \frac{AB}{\frac{1}{2}AB + AB} = \frac{2}{3},$$

since $CD' = AB$, because of the fact that triangles MCD' and MBA are congruent (A.S.A.). But from this equation we get $\frac{BP}{NP+BP} = \frac{2}{3+2}$, i.e., $\frac{BP}{BN} = \frac{2}{5}$. In the same way, since triangles BPM and PAA' are similar, we have

$$\frac{MP}{AP} = \frac{BP}{PA'} = \frac{MB}{AA'} = \frac{MB}{2AD} = \frac{\frac{1}{2}AD}{2AD} = \frac{1}{4}.$$

(Here we used the fact that $AD = DA'$, since triangles $A'DN$ and BNC are congruent.) Therefore $\frac{MP+AP}{AP} = \frac{1+4}{4}$, i.e., $\frac{AP}{AM} = \frac{4}{5}$.

S2005-8 We write $x = 5k + x_1$, $y = 5l + y_1$, and $z = 5m + z_1$, where k, l, m, x_1, y_1, z_1 are integers, and without loss of generality we can assume that $0 \leq x_1 \leq y_1 \leq z_1 < 5$. If 25 divides $x^5 + y^5 + z^5$, then using the Binomial Theorem we see that 25 must also divide $x_1^5 + y_1^5 + z_1^5$. We will show that $x_1 = 0$, which implies 25 divides $y^5 + z^5$. It follows from expanding the polynomial $(x_1 + y_1 + z_1)^5$ that since 5 divides $x_1^5 + y_1^5 + z_1^5$, 5 also divides $(x_1 + y_1 + z_1)^5$. For any integer a, we have $a^5 \equiv a \pmod 5$ by Fermat's Little Theorem (see [HWr] or [M]), and so 5 must divide $x_1 + y_1 + z_1$. Toward a contradiction, suppose $x_1 > 0$. Then $3 \leq x_1 + y_1 + z_1 \leq 12$, and there are only four choices for the triple (x_1, y_1, z_1) such that 5 divides $x_1 + y_1 + z_1$, namely

$$(1, 1, 3), (1, 2, 2), (2, 4, 4), \text{ and } (3, 3, 4).$$

We have

$$1^5 \equiv 1 \pmod{25},$$

$$2^5 \equiv 7 \pmod{25},$$
$$3^5 \equiv -7 \pmod{25},$$
$$4^5 \equiv -1 \pmod{25}.$$

By inspection, for none of the four choices above is the sum $x_1^5 + y_1^5 + z_1^5$ divisible by 25, which represents a contradiction. Therefore, $x_1 = 0$, and the conclusion follows.

Chapter 29
2006 Solutions

The 41st competition was held at Taylor University for the first time.

The **2006 problem set** was prepared by Professor Mike Axtell at Wabash College and Professor Joe Stickles at the University of Evansville.

The **2006 problem statements** begin on page 13 in Chap. 6.

S2006-1 Rewriting, we notice that

$$f(x) = \sqrt{16 - (x-2)^2} - \sqrt{1 - (x-5)^2}.$$

So, the values of $f(x)$ are the differences of the y coordinates of two semicircles, and these semicircles are internally tangent at the point $(6, 0)$. Furthermore, we notice that the domain for $f(x)$ is $[4, 6]$. Since the function

$$g(x) = \sqrt{16 - (x-2)^2}$$

is decreasing on $[4, 6]$, the expression $\sqrt{4x - x^2 + 12}$ achieves its maximum value at $x = 4$, and at this point $g(4) = \sqrt{4(4) - 4^2 + 12} = 2\sqrt{3}$. Furthermore, at $x = 4$, the function $h(x) = \sqrt{1 - (x-5)^2}$ is as small as it can get; namely, $h(4) = 0$. Hence the maximum value of $f(x) = g(x) - h(x)$ is $f(4) = 2\sqrt{3}$.

S2006-2 Using the hint, we see for $n \geq 1$,

$$\arctan(1/n) - \arctan(1/(n+2)) = \arctan\left(\frac{\frac{1}{n} - \frac{1}{n+2}}{1 + \frac{1}{n(n+2)}}\right)$$

$$= \arctan(2/(n+1)^2).$$

Let S_N denote the Nth partial sum of the given series. For $N \geq 3$,

$$S_N = \sum_{n=1}^{N} \arctan\left(\frac{2}{n^2}\right) = \arctan 2 + \sum_{n=2}^{N} \arctan\left(\frac{2}{n^2}\right)$$

$$= \arctan 2 + \sum_{n=1}^{N-1} \arctan\left(\frac{2}{(n+1)^2}\right)$$

$$= \arctan 2 + \sum_{n=1}^{N-1} \left(\arctan\left(\frac{1}{n}\right) - \arctan\left(\frac{1}{n+2}\right)\right)$$

$$= \arctan 2 + \arctan 1 + \arctan\left(\frac{1}{2}\right) - \arctan\left(\frac{1}{N}\right) - \arctan\left(\frac{1}{N+1}\right).$$

As $N \to \infty$, we have $S_N \to \arctan 2 + \arctan 1 + \arctan(\frac{1}{2}) = \frac{3\pi}{4}$. The last equality holds because $\arctan 1 = \frac{\pi}{4}$, and for $x > 0$, we have $\arctan(x) + \arctan(\frac{1}{x}) = \frac{\pi}{2}$.

S2006-3 In order for $\left[\sqrt{10X}\right] = 20$, we need $20 \leq \sqrt{10X} < 21$ or $40 \leq X < 44.1$. Given $\left[\sqrt[3]{X}\right] = 3$, we know $3 \leq \sqrt[3]{X} < 4$ or $27 \leq X < 64$. So, we find

$$P\left(\left[\sqrt{10X}\right] = 20 \mid \left[\sqrt[3]{X}\right] = 3\right) = \left(\int_{40}^{44.1} \frac{1}{x^2} dx\right) / \left(\int_{27}^{64} \frac{1}{x^2} dx\right)$$

$$= \left(\frac{1}{40} - \frac{1}{44.1}\right) / \left(\frac{1}{27} - \frac{1}{64}\right) = \frac{984}{9065}.$$

S2006-4 Modeling the position x by the continuous, increasing function $f(t)$, for time $0 \leq t \leq 72$, with $f(0) = 0$ and $f(72) = 18$, the inverse function $t = f^{-1}(x)$ is continuous and increasing for $0 \leq x \leq 18$ (the continuity of the inverse is a well-known fact from calculus which is not proved here). Let $T(x)$ be the time in minutes that it takes the rider to go from point x to point $x + 3$. Certainly, $T(x) = f^{-1}(x+3) - f^{-1}(x)$ is continuous on $[0, 15]$. Notice that

$$T(0) + T(3) + T(6) + T(9) + T(12) + T(15) = 72.$$

It is not possible for all of the values on the left-hand side to be strictly greater than 12, nor is it possible for all of the values on the left-hand side of the equation to be strictly less than 12. Hence, there exist $r, s \in \{0, 3, 6, 9, 12, 15\}$ with $T(r) \leq 12$ and $T(s) \geq 12$. Since $T(x)$ is continuous, the Intermediate Value Theorem guarantees that there exists a k between r and s (or $k = r$ or $k = s$) so that $T(k) = 12$. In other words, there exists a point k so that the rider travels from point k to point $k + 3$ in exactly 12 minutes.

Remark from the Editors This solution uses, in an essential way [W], the fact that 18 miles is an integer multiple of 3 miles.

S2006-5 Toward a contradiction, assume that neither T nor U is closed under multiplication. Thus, there exist $t_1, t_2 \in T$ such that $t_1 t_2 \notin T$ and $u_1, u_2 \in U$ such that $u_1 u_2 \notin U$. Since $S = T \cup U$, we have $t_1 t_2 \in U$ and $u_1 u_2 \in T$. Consider the element $t_1 t_2 u_1 u_2$. Since the product of any three elements of T (respectively, U) is in T (respectively U); we have $t_1 t_2 u_1 u_2 = (t_1 t_2) u_1 u_2 \in U$ and $t_1 t_2 u_1 u_2 = t_1 t_2 (u_1 u_2) \in T$. So, $t_1 t_2 u_1 u_2 \in T \cap U$. However, T and U are disjoint, a contradiction. Thus, either T or U is closed under multiplication.

S2006-6 Suppose x and y are integers satisfying the given equation. Factoring the left-hand side of the equation, we see

$$((r-s)x - (r+s)y)((r+s)x + (r-s)y) = 1.$$

Since r, s, x, y are integers, we have the product of two integers, and this product must be 1. Thus, the factors in the product are either both 1 or both -1. Solving the system $(r-s)x - (r+s)y = \delta$, $(r+s)x + (r-s)y = \delta$, where $\delta = \pm 1$, we obtain $x = \frac{r\delta}{r^2+s^2}$ and $y = \frac{-s\delta}{r^2+s^2}$. For nonzero integers r and s, these solutions satisfy $0 < |x| < 1$ and $0 < |y| < 1$, so neither x nor y is an integer, a contradiction.

S2006-7

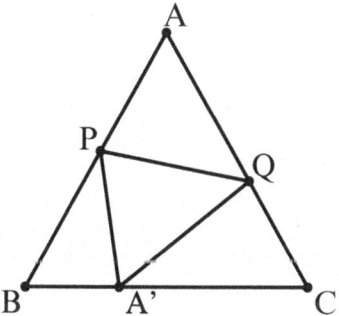

Let $x = PA = PA'$ and $y = QA = QA'$. Then $PB = 3 - x$ and $QC = 3 - y$. Applying the Law of Cosines to $\triangle PBA'$, we have $x^2 = 1^2 + (3-x)^2 - 2(1)(3-x)\cos 60°$. Solving for x, we get $x = \frac{7}{5}$. Applying the Law of Cosines to $\triangle QCA'$, we have $y^2 = 2^2 + (3-y)^2 - 2(2)(3-y)\cos 60°$. Solving for y, we get $y = \frac{7}{4}$. Finally, applying the Law of Cosines to $\triangle PA'Q$, we have $PQ^2 = (\frac{7}{5})^2 + (\frac{7}{4})^2 - 2(\frac{7}{5})(\frac{7}{4})\cos 60°$. Solving for PQ, we get $PQ = \frac{7\sqrt{21}}{20}$.

S2006-8 Label the twelve stones $\{A, B, C, D, E, F, G, H, I, J, K, L\}$. Weighing 1 is $\{A, B, C, D\}$ versus $\{E, F, G, H\}$. If the scales balance, go to 1. If the scales do not balance, go to 2.

1. The scales balanced, meaning the odd stone is in $\{I, J, K, L\}$. Weighing 2 is $\{I, J, K\}$ versus the good stones $\{A, B, C\}$. If the scales balance, go to 1a. If the scales do not balance, go to 1b.
1a. The odd stone must be L. Weighing L against any other stone will determine whether L is heavier or lighter than the others.
1b. The odd stone is in $\{I, J, K\}$, and it is light (respectively heavy) because $\{I, J, K\}$ is lighter (respectively heavier) than $\{A, B, C\}$. Weighing 3 is I versus J. If they balance, then K is the odd stone and light (respectively heavy). If they do not balance, then the light (respectively heavy) side is the odd stone.
2. The scales tipped. Without loss of generality, assume that $\{A, B, C, D\}$ was the heavier side. So, we either have an odd, heavy stone among $\{A, B, C, D\}$ or an odd, light stone among $\{E, F, G, H\}$. Weighing 2 is $\{A, B, E\}$ versus $\{F, C, D\}$. If they balance, go to 2a. If $\{A, B, E\}$ is heavier than $\{F, C, D\}$, go to 2b. If $\{A, B, E\}$ is lighter than $\{F, C, D\}$, go to 2c.
2a. The odd stone must be in $\{G, H\}$, and it must be light since $\{G, H\}$ were on the light side of the first measurement. Measure G against H; the light side is the odd, light stone.
2b. $\{A, B, E\}$ was heavier than $\{F, C, D\}$. Neither C nor D can be an odd, heavy stone, and E cannot be an odd, light stone. Thus, either A or B is the odd, heavy stone, or F is the odd, light stone. Weighing 3 is $\{A, F\}$ against any two normal stones (e.g., $\{G, H\}$). If $\{A, F\}$ is heavier, then A is the odd, heavy stone. If $\{A, F\}$ is lighter, then $\{F\}$ is the odd, light stone. If they balance, then $\{B\}$ is the odd, heavy stone.
2c. $\{A, B, E\}$ was lighter than $\{F, C, D\}$. Neither A nor B can be an odd, heavy, while F cannot be odd and light. Thus, either C or D is the odd, heavy stone, or E is the odd, light stone. Weighing 3 is $\{C, E\}$ against any two normal stones (e.g., $\{G, H\}$). If $\{C, E\}$ is heavy, then C is the odd, heavy stone. If $\{C, E\}$ is light, then E is the odd, light stone. If they balance, then D is the odd, heavy stone.

Remark from the Editors Coin balancing problems like this are well-known; for a survey paper, we refer to [GN]. This version with 12 coins is due to H. Grossman [G].

Chapter 30
2007 Solutions

The 42nd competition was held at the University of Indianapolis.

The 2007 problem set was prepared by Professor Mike Axtell at Wabash College and Professor Joe Stickles at Millikin University.

The **2007 problem statements** begin on page 15 in Chap. 7.

S2007-1 If $x = \sqrt{p} + \sqrt{q}$, then $x - \sqrt{p} = \sqrt{q}$. Squaring both sides, we obtain $x^2 - 2x\sqrt{p} + p = q$, or $-2x\sqrt{p} = q - p - x^2 = -x^2 + (q - p)$. Square both sides again, we get $4px^2 = x^4 - 2(q-p)x^2 + (q-p)^2$, so $f(x) = x^4 - 2(q+p)x^2 + (q-p)^2$ will suffice.

S2007-2 We know that $\gcd(36, n) = x$ and $\text{lcm}(36, n) = 500 + x$. Now, since x is a divisor of 36, we have the following possibilities for x: 1, 2, 3, 4, 6, 9, 12, 18, 36. Now, 36 must also divide $500 + x$, and by experimentation we get $x = 4$. This tells us that $n = 4b$ and that $4b$ must divide 504. Hence, b must divide 126. However, $n = 4b$ cannot have 3 as a factor, and since $126 = 2 \cdot 3^2 \cdot 7$, the possible values of b are 1, 2, 7, and 14. We quickly see 14 is the only possibility. Hence $n = 4 \cdot 2 \cdot 7 = 56$.

S2007-3 For $x \leq t < 3x$, we have $0 < \dfrac{1}{t\sqrt{t^4+1}} \leq \dfrac{1}{x\sqrt{x^4+1}}$. Hence

$$0 < \int_x^{3x} \frac{dt}{t\sqrt{t^4+1}} \leq \int_x^{3x} \frac{dt}{x\sqrt{x^4+1}}$$

$$= \frac{1}{x\sqrt{x^4+1}} \int_x^{3x} dt = \frac{1}{x\sqrt{x^4+1}} \cdot (2x) = \frac{2}{\sqrt{x^4+1}}.$$

Thus

$$0 \leq \lim_{x \to \infty} (x+2) \cdot \int_x^{3x} \frac{dt}{t\sqrt{t^4+1}} \leq \lim_{x \to \infty} (x+2) \cdot \frac{2}{\sqrt{x^4+1}} = 0.$$

So, $\lim_{x \to \infty} (x+2) \cdot \int_x^{3x} \frac{dt}{t\sqrt{t^4+1}} = 0.$

S2007-4

(a) Reduce the integers modulo p, obtaining a uniform distribution over the ordered list of Np remainders $(1, 2, \ldots, p-1, 0, 1, 2, \ldots, p-1, 0, 1, 2, \ldots, p-1, 0)$. For any N, the probability a is divisible by p is the same as the probability of selecting 0 from $(1, 2, \ldots, p-1, 0)$, which is $\frac{1}{p}$.

(b) The selection of a and the selection of b are independent events. So, using part (a), we see the probability is $\frac{1}{p} \cdot \frac{1}{p} = \frac{1}{p^2}$.

(c) Part (b) applies: $K = 2 \cdot 3 \cdots p_k$ is of the form Np for any of the first k prime numbers p. For each such prime p, the probability a and b both have p as a factor is $\frac{1}{p^2}$; hence, the probability a and b are not both divisible by p is $1 - \frac{1}{p^2}$. Now, if p_1 and p_2 are distinct primes, whether or not p_1 divides both a and b is independent from whether or not p_2 divides a and b. Hence, the probability that neither p_1 nor p_2 divides both a and b is $\left(1 - \frac{1}{p_1^2}\right)\left(1 - \frac{1}{p_2^2}\right)$. Continuing, we see that the probability that a and b have no common factor from among the first k primes is $P(k) = \left(1 - \frac{1}{2^2}\right)\left(1 - \frac{1}{3^2}\right) \cdots \left(1 - \frac{1}{p_k^2}\right)$. For sufficiently large k, this is arbitrarily close to both the $k \to \infty$ limit and the product over all primes, $\prod_{p \in \mathcal{P}} \left(1 - \frac{1}{p^2}\right)$.

Remark from the Editors This infinite product is known to converge to $\frac{6}{\pi^2}$ (see [HWr]).

S2007-5 First, note that $A = B - I$, where I is the $n \times n$ identity matrix, and that $B^2 = nB$. For any real number r, we see $(B - I)(rB - I) = rB^2 - (r+1)B + I = (rn - (r+1))B + I$, so $rB - I$ will be the inverse of $B - I$ if $rn - (r+1) = 0$, or $r = \frac{1}{n-1}$. Hence,

$$A^{-1} = \frac{1}{n-1} B - I = \begin{bmatrix} \frac{2-n}{n-1} & \frac{1}{n-1} & \cdots & \frac{1}{n-1} \\ \frac{1}{n-1} & \frac{2-n}{n-1} & \cdots & \frac{1}{n-1} \\ \vdots & \vdots & \ddots & \vdots \\ \frac{1}{n-1} & \frac{1}{n-1} & \cdots & \frac{2-n}{n-1} \end{bmatrix}.$$

S2007-6 Note that $h^9 = (ghg^{-1})^3 = gh^3g^{-1} = g(ghg^{-1})g^{-1}$. So $h^{27} = (ghg^{-1})^9 = gh^9g^{-1} = g(g^2hg^{-2})g^{-1} = g^3hg^{-3}$. Since $g^3 = e$, $h^{27} = h$, or $h^{26} = e$. Since the group has odd order, the only possibilities for the order of h are 1 and 13. Since g and h do not commute, $h \neq e$; hence $|h| = 13$.

Chapter 31
2008 Solutions

The 43rd competition was held at Saint Mary's College.

The 2008 problem set was prepared by Professor Mike Axtell at Wabash College and Professor Joe Stickles at Millikin University.

The **2008 problem statements** begin on page 17 in Chap. 8.

S2008-1 We have

$$yx = (yx)^3 = (yx)^2(yx) = ((yx)^2 y)x$$
$$= y(yx)^2 x = y(yx)(yx)x$$
$$= xy^2 yx^2 = xyx^2$$
$$= xx^2 y = x^3 y = xy.$$

Remark from the problem authors This problem and solution were both adapted from ICMC **P1974-2** [AFMC].

S2008-2 Consider the set $[0, 4] \times [0, 4]$. Draw in the line $y = x$, which represents both students meeting each other at the exact same time. If x arrives at time t, then y can arrive at any time between $t - 0.5$ and $t + 0.5$, and the students will meet. Similarly, if y arrives at time t', then x can arrive at any time between $t' - 0.5$ and $t' + 0.5$, and the students will meet. So, widening the line $y = x$ so that it is everywhere 1 hour wide and 1 hour tall forms a region that represents all possible "successful" meeting times. The area of the 4×4-square represents all possible selections of times by both students.

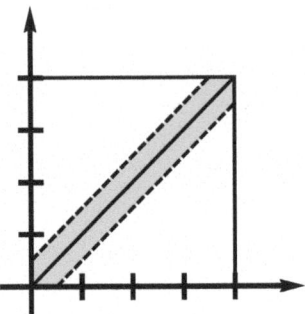

The area of the shaded region is $16 - 2(\frac{1}{2})(\frac{7}{2})^2 = \frac{15}{4}$. So, the probability the two students meet during this 4-hour period is $\frac{15/4}{16} = \frac{15}{64}$.

S2008-3 Let the angles of the triangles be denoted by α, β, γ and α', β', γ', and let $\alpha = \alpha'$. We ask under what conditions the following inequality holds: $\sin \alpha + \sin \beta + \sin \gamma < \sin \alpha' + \sin \beta' + \sin \gamma'$, or since $\sin \alpha = \sin \alpha'$, $\sin \beta + \sin \gamma < \sin \beta' + \sin \gamma'$. By the hint, this inequality can be written as $2\sin(\frac{\beta+\gamma}{2})\cos(\frac{\beta-\gamma}{2}) < 2\sin(\frac{\beta'+\gamma'}{2})\cos(\frac{\beta'-\gamma'}{2})$. Since $\alpha = \alpha'$, we have $\beta + \gamma = \beta' + \gamma' < 180°$, so that $2\sin(\frac{\beta+\gamma}{2}) = 2\sin(\frac{\beta'+\gamma'}{2}) > 0$. Dividing both sides by $2\sin(\frac{\beta+\gamma}{2})$ yields $\cos(\frac{\beta-\gamma}{2}) < \cos(\frac{\beta'-\gamma'}{2})$. Since $\cos\theta = \cos|\theta|$ and $y = \cos\theta$ is decreasing on $0° \le \theta \le 180°$, the above inequality holds if and only if $\left|\frac{\beta'-\gamma'}{2}\right| < \left|\frac{\beta-\gamma}{2}\right|$, or equivalently, $|\beta' - \gamma'| < |\beta - \gamma|$.

Remark from the problem authors This problem and solution were both adapted from Problem #2 of the 1898 Eötvös Competition in Hungary [K₁].

S2008-4 The cross-sectional area is $\frac{1}{2}\pi(\sqrt{2y_i})^2 = \pi y_i$ at level y_i. The volume of one puffed-up cross-section is $\pi y_i \Delta y_i$, and we sum these together to get an approximate volume of $\sum \pi y_i \Delta y_i$. This quickly leads to the exact volume given by $\int_0^2 \pi y \, dy = 2\pi$.

S2008-5 First, we notice that

$$x_n = \frac{x_{n-1} + (n-1)x_{n-2}}{n}$$
$$= \frac{nx_{n-1} - (n-1)x_{n-1} + (n-1)x_{n-2}}{n}$$
$$= x_{n-1} + \frac{-(n-1)x_{n-1} + (n-1)x_{n-2}}{n},$$

or $x_n - x_{n-1} = -\frac{n-1}{n}(x_{n-1} - x_{n-2})$. Similar arguments show $x_{n-1} - x_{n-2} = -\frac{n-2}{n-1}(x_{n-2}-x_{n-3})$, $x_{n-2}-x_{n-3} = -\frac{n-3}{n-2}(x_{n-3}-x_{n-4}), \ldots, x_2-x_1 = -\frac{1}{2}(x_1-x_0)$.
Substituting, we see for $n \geq 2$,

$$x_n - x_{n-1} = \left(-\frac{n-1}{n}\right) \cdot \left(-\frac{n-2}{n-1}\right) \cdot \left(-\frac{n-3}{n-2}\right) \cdot \ldots \cdot \left(-\frac{1}{2}\right)(x_1 - x_0)$$

$$= \frac{(-1)^{n-1}}{n}.$$

We also notice that the above equation is true when $n = 1$. Since $x_0 = 0$, we can write

$$\begin{aligned}
x_n &= x_n - x_0 \\
&= (x_n - x_{n-1}) + (x_{n-1} - x_{n-2}) + \cdots + (x_1 - x_0) \\
&= \sum_{k=1}^{n}(x_k - x_{k-1}) \\
&= \sum_{k=1}^{n} \frac{(-1)^{k-1}}{k} \\
&= \sum_{k=0}^{n-1} \frac{(-1)^k}{k+1}.
\end{aligned}$$

Hence $\lim_{n \to \infty} x_n = \sum_{k=0}^{\infty} \frac{(-1)^k}{k+1} = \ln 2$.

S2008-6 We have

$$\begin{aligned}
\frac{1}{n+1}\binom{2n}{n} &= \frac{1}{n+1} \cdot \frac{(2n)!}{(n!)^2} = \frac{(2n)!}{(n+1)!n!} \\
&= \frac{(2n)!(2n+1-2n)}{(n+1)!n!} = \frac{(2n)!(2n+1)}{(n+1)!n!} - \frac{(2n)!(2n)}{(n+1)!n!} \\
&= \frac{(2n+1)!}{(n+1)!n!} - \frac{2n}{n} \cdot \frac{(2n)!}{(n+1)!(n-1)!} \\
&= \frac{(2n+1)!}{(n+1)!((2n+1)-(n+1))!} - 2 \cdot \frac{(2n)!}{(n+1)!(2n-(n+1))!} \\
&= \binom{2n+1}{n+1} - 2\binom{2n}{n+1}.
\end{aligned}$$

As both $\binom{2n+1}{n+1}$ and $\binom{2n}{n+1}$ are integers for all integers $n \geq 1$, we have $\frac{1}{n+1}\binom{2n}{n}$ is an integer for all integers $n \geq 1$ as well.

S2008-7A Let $A = \begin{bmatrix} 1 & -1 \\ 0 & 5 \end{bmatrix}$. Then the characteristic polynomial of A is $\lambda^2 - 6\lambda + 5$. By the Cayley–Hamilton Theorem, we have $A^2 - 6A + 5I = 0$. Multiplying both sides by A, we obtain $A^3 - 6A^2 + 5A = 0$. Now consider $(A - 2I)^3$. We have $(A - 2I)^3 = A^3 - 6A^2 + 12A - 8I = A^3 - 6A^2 + 5A + 7A - 8I = 7A - 8I$. So, we have $7A = (A - 2I)^3 + 8I$, or $A = \left(\frac{1}{\sqrt[3]{7}}(A - 2I)\right)^3 + \left(\frac{2}{\sqrt[3]{7}}I\right)^3$. Thus, let

$$B = \frac{1}{\sqrt[3]{7}}(A - 2I) = \begin{bmatrix} -\frac{1}{\sqrt[3]{7}} & -\frac{1}{\sqrt[3]{7}} \\ 0 & \frac{3}{\sqrt[3]{7}} \end{bmatrix} \text{ and } C = \frac{2}{\sqrt[3]{7}}I = \begin{bmatrix} \frac{2}{\sqrt[3]{7}} & 0 \\ 0 & \frac{2}{\sqrt[3]{7}} \end{bmatrix}.$$

S2008-7B The sum $B^3 + C^3$ is upper triangular, so try the upper triangular matrices $B = \begin{bmatrix} a & b \\ 0 & c \end{bmatrix}$ and $C = \begin{bmatrix} 0 & 0 \\ 0 & 0 \end{bmatrix}$. Expand to get

$$B^3 + C^3 = B^3 = \begin{bmatrix} a^3 & a^2b + abc + bc^2 \\ 0 & c^3 \end{bmatrix} = \begin{bmatrix} 1 & -1 \\ 0 & 5 \end{bmatrix},$$

and let $a = 1$, $c = 5^{1/3}$. Equating the upper right entries

$$b + 5^{1/3}b + 5^{2/3}b = -1,$$

there is a solution for b, which gives an example of the required matrices:

$$B = \begin{bmatrix} 1 & -1/\left(1 + 5^{1/3} + 5^{2/3}\right) \\ 0 & 5^{1/3} \end{bmatrix}, \quad C = \begin{bmatrix} 0 & 0 \\ 0 & 0 \end{bmatrix}.$$

Chapter 32

2009 Solutions

The 44th competition was held at Indiana University—Purdue University Indianapolis.

The 2009 problem set was prepared by Professor Chris Mitchell at the host institution.

The **2009 problem statements** begin on page 19 in Chap. 9.

S2009-1 The area between the lines $P = Q + \frac{1}{3}$ and $P = Q - \frac{1}{3}$ that is contained in the square $[0, 2] \times [0, 2]$ is equal to $4 - (\frac{5}{3})^2 = \frac{11}{9}$. Divide this number by 4 to get the required probability.

S2009-2 Consider the equation

$$\frac{x}{1 - x - x^2} = \sum_{n=1}^{\infty} f_n x^n.$$

Multiply both sides of this equation by the $1 - x - x^2$, and expand the right-hand side of the equation. The result, after reindexing, is the equation

$$x = f_1 x + (f_2 - f_1)x^2 + \sum_{n=3}^{\infty}(f_n - f_{n-1} - f_{n-2})x^n.$$

The result follows by comparing coefficients on both sides of the equation.

S2009-3 Note that, using cycle notation, $\sigma^3 = (1\ 2\ 3\ 5\ 7\ 8\ 4\ 6)$. Because σ^3 can be written as a single 8-cycle, σ must also be writable as some 8-cycle (by the uniqueness of expressing permutations as a product of disjoint cycles). Because all 8-cycles have an order of 8, it follows that $\sigma = \sigma^9 = (\sigma^3)^3 = (1\ 5\ 4\ 2\ 7\ 6\ 3\ 8)$.

S2009-4 Let $v = \sqrt[3]{2+\sqrt{5}} + \sqrt[3]{2-\sqrt{5}}$. Note that $v \in \mathbb{R}$, and expanding v^3 shows that v satisfies the equation $x^3 + 3x - 4 = 0$. Since $x^3 + 3x - 4 = (x-1)(x^2+x+4)$ and the discriminant of $x^2 + x + 4$ is -15, the only real solution of $x^3 + 3x - 4 = 0$ is 1. So, $v = 1$.

Remark from the Editors This problem is similar to **P2014-8**—the solution **S2014-8** in Chap. 37 shows a more detailed calculation.

S2009-5 A stronger result holds under the given conditions: $n \mid (n-3)!$. A composite number $0 < n \neq 4$ satisfies $n \geq 6$ and $n = a \cdot b$ for $1 < a \leq b \leq \frac{1}{2}n \leq n - 3$. If $a < b$, then a and b are distinct factors of $(n-3)!$, so $n = ab$ divides $(n-3)!$. If $a = b$, then $2 < a < 2a \leq (a-1)a = a^2 - a \leq n - 3$, so a and $2a$ are distinct factors of $(n-3)!$ and $n = a^2$ divides $(n-3)!$.

Remark from the Editors This well-known problem, showing that Wilson's Theorem only works for primes, is considered, for example, in [RL].

S2009-6A Step 1. For any three points in the interval, $a < r < s < t < b$, the inequality

$$f(\lambda x + (1-\lambda)y) \leq \lambda f(x) + (1-\lambda)f(y)$$

applied to $x = r, y = t, \lambda = \dfrac{t-s}{t-r} \in (0,1)$ gives

$$f\left(\frac{t-s}{t-r}r + \left(1 - \frac{t-s}{t-r}\right)t\right) \leq \frac{t-s}{t-r}f(r) + \left(1 - \frac{t-s}{t-r}\right)f(t)$$

$$\implies f(s) \leq \frac{t-s}{t-r}f(r) + \frac{s-r}{t-r}f(t).$$

This implies (without assuming f is increasing)

$$\frac{f(s) - f(r)}{s-r} \leq \left(\left(\frac{t-s}{t-r}f(r) + \frac{s-r}{t-r}f(t)\right) - f(r)\right)/(s-r)$$

$$= \frac{f(t) - f(r)}{t-r}$$

$$= \left(f(t) - \left(\frac{t-s}{t-r}f(r) + \frac{s-r}{t-r}f(t)\right)\right)/(t-s)$$

$$\leq \frac{f(t) - f(s)}{t-s}.$$

Step 2. To show that f is continuous at a particular point $x_0 \in (a, b)$, let $\epsilon > 0$, and pick any point $t_0 \in (x_0, b)$. Let $m = \dfrac{f(t_0) - f(x_0)}{t_0 - x_0} > 0$ (using f increasing), and let $\delta = \min\{\epsilon/(2m), t_0 - x_0, x_0 - a\} > 0$. If $0 < x - x_0 < \delta$, then $a < x_0 < x < t_0 < b$ and, using Step 1,

$$0 < f(x) - f(x_0) = \frac{f(x) - f(x_0)}{x - x_0} \cdot (x - x_0)$$
$$\leq \frac{f(t_0) - f(x_0)}{t_0 - x_0} \cdot \delta \leq m \cdot (\epsilon/(2m)) < \epsilon.$$

If $0 < x_0 - x < \delta$, then $a < x < x_0 < t_0 < b$, and using Step 1,

$$0 < f(x_0) - f(x) = \frac{f(x_0) - f(x)}{x_0 - x} \cdot (x_0 - x)$$
$$\leq \frac{f(t_0) - f(x_0)}{t_0 - x_0} \cdot \delta \leq m \cdot (\epsilon/(2m)) < \epsilon.$$

The conclusion is that $|f(x) - f(x_0)| < \epsilon$, so $\lim\limits_{x \to x_0} f(x) = f(x_0)$ and f is continuous.

S2009-6B A stronger conclusion is possible, even without the increasing assumption: Both one-side derivatives of f exist at each point (but are not necessarily equal, e.g., $f(x) = |x|$).

To show that f has a right-side derivative at a particular point $x_0 \in (a, b)$, consider the set

$$\left\{ \frac{f(x) - f(x_0)}{x - x_0} : x_0 < x < b \right\}.$$

This set is non-empty and bounded below: For some $r_0 \in (a, x_0)$,

$$\frac{f(x) - f(x_0)}{x - x_0} \geq \frac{f(x_0) - f(r_0)}{x_0 - r_0}$$

by Step 1. So, the set has a greatest lower bound, L, and for any $\epsilon > 0$, there is some $t_0 \in (x_0, b)$ such that

$$L \leq \frac{f(t_0) - f(x_0)}{t_0 - x_0} < L + \epsilon.$$

If $0 < x - x_0 < t_0 - x_0$, then by Step 1,

$$L \leq \frac{f(x) - f(x_0)}{x - x_0} \leq \frac{f(t_0) - f(x_0)}{t_0 - x_0} < L + \epsilon.$$

This shows
$$\lim_{x \to x_0^+} \frac{f(x) - f(x_0)}{x - x_0} = L = f'_+(x_0).$$

The right-side continuity follows from the usual calculation:

$$\lim_{x \to x_0^+} (f(x) - f(x_0)) = \lim_{x \to x_0^+} \left(\frac{f(x) - f(x_0)}{x - x_0} \cdot (x - x_0) \right) = f'_+(x_0) \cdot 0 = 0.$$

In the same way, the left-side derivative $f'_-(x_0)$ is the least upper bound of the set of left-side difference quotients. The left-side continuity and the two-side continuity follow.

S2009-7A There may be many such bijections from $(\mathbb{Z}_{\geq 0})^3$ to $\mathbb{Z}_{\geq 1}$—the following example is constructed using the increasing sequence of triangular numbers

$$(0, 1, 3, 6, 10, \ldots, T_a, \ldots),$$

defined by $T_a = \frac{1}{2}a(a+1)$, or recursively by $T_0 = 0$, and $T_{a+1} = T_a + a + 1$, and another increasing sequence, the tetrahedral numbers,

$$(0, 1, 4, 10, 20, 35, 56, 84, 120, 165, 220, 286, 364, 455,$$
$$560, 680, 816, 969, 1140, 1330, 1540, 1771, 2024, \ldots, E_b, \ldots),$$

defined by $E_b = \frac{1}{6}b(b+1)(b+2)$, or recursively by $E_0 = 0$ and $E_{b+1} = E_b + T_{b+1}$. Then define $F : (\mathbb{Z}_{\geq 0})^3 \to \mathbb{Z}_{\geq 1}$ by

$$F(p, q, r) = 1 + p + T_{p+q} + E_{p+q+r}.$$

To show F is surjective, consider $x \geq 1$, and let $b = \max\{y \in \mathbb{Z} : E_y < x\}$, so that $E_b < x \leq E_{b+1}$ and $0 < x - E_b \leq E_{b+1} - E_b = T_{b+1}$. Let $a = \max\{y \in \mathbb{Z} : T_y < x - E_b\}$, so that $T_a < x - E_b \leq T_{a+1}$ and $0 < x - E_b - T_a \leq T_{a+1} - T_a = a + 1$. From $T_a < x - E_b \leq T_{b+1}$, we can conclude $a < b + 1$, so $b - a \geq 0$. Then one can verify that

$$F(x - E_b - T_a - 1, a - (x - E_b - T_a - 1), b - a) = x.$$

To show F is injective, suppose there are two input triples; label the pair with the smaller or equal sum of components by (p, q, r) and the other pair by (p^*, q^*, r^*), so $p + q + r \leq p^* + q^* + r^*$.

We want to show that assuming the outputs are equal

$$F(p,q,r) = F(p^*, q^*, r^*)$$
$$1 + p + T_{p+q} + E_{p+q+r} = 1 + p^* + T_{p^*+q^*} + E_{p^*+q^*+r^*} \qquad (32.1)$$

leads to the conclusion $p = p^*$ and $q = q^*$ and $r = r^*$.

Case 1. $p+q+r < p^* + q^* + r^*$. Then, using the increasing property of the E_b and T_a sequences, (32.1) implies

$$\begin{aligned} 0 &= E_{p^*+q^*+r^*} - E_{p+q+r} + T_{p^*+q^*} - T_{p+q} + p^* - p \\ &\geq E_{p+q+r+1} - E_{p+q+r} - T_{p+q} + p^* - p \\ &= T_{p+q+r+1} - T_{p+q} + p^* - p \\ &\geq T_{p+q+1} - T_{p+q} + p^* - p \\ &= p + q + 1 + p^* - p \\ &= p^* + q + 1 \geq 1, \end{aligned}$$

a contradiction.

Case 2. $p+q+r = p^* + q^* + r^*$ and $p+q < p^* + q^*$. From (32.1),

$$\begin{aligned} 0 &= T_{p^*+q^*} - T_{p+q} + p^* - p \\ &\geq T_{p+q+1} - T_{p+q} + p^* - p \\ &= p + q + 1 + p^* - p \\ &= p^* + q + 1 \geq 1, \end{aligned}$$

a contradiction.

Case 3. $p+q+r = p^* + q^* + r^*$ and $p+q > p^* + q^*$. Then (32.1) leads to a contradiction in the same way as Case 2.

Case 4. The only remaining case is $p+q+r = p^*+q^*+r^*$ and $p+q = p^*+q^*$, so $r = r^*$ follows immediately, and then $p = p^*$ follows from (32.1), and so in this case, $q = q^*$.

Remark from the Editors There may be descriptions, sketches, or algorithms describing such a correspondence, generalizing the better-known diagonal construction of a map $(\mathbb{Z}_{\geq 0})^2$ to $\mathbb{Z}_{\geq 1}$ from two dimensions to three, but the above construction gives a concrete formula as stated in the problem.

S2009-7B If an explicit formula is known for a bijection $g : (\mathbb{Z}_{\geq 0})^2 \to \mathbb{Z}_{\geq 1}$ (as in **P2022-8**, see the solution **S2022-8** in Chap. 44), then $F(p,q,r) = g(g(p,q)-1, r)$ is a bijection because it is a composite of bijections.

Chapter 33
2010 Solutions

The 45th competition was held at Franklin College.

The 2010 problem set was prepared by Professor Chris Mitchell at American University.

The **2010 problem statements** begin on page 21 in Chap. 10.

S2010-1 The solution is $\binom{p+q}{q}$. In order to see this, think of $p+q$ boxes. Each box will contain a card. The deck is divided into two parts, one containing p cards and one containing q cards. Shuffling the two parts together is the same as selecting q of the $p+q$ boxes and then placing the cards in order into those boxes.

S2010-2 First, recall Wilson's Theorem (see [HWr], [M], or [RL]):

$$(p-1)! \equiv -1 \pmod{p}.$$

Note that for every $0 < i \leq k$, $p - i \equiv -i \pmod{p}$. As a result

$$(p-(k+1))!k! \equiv (-1)^k (p-(k+1))!(p-k)\cdots(p-1) \pmod{p}.$$

S2010-3 Player 1 should take 1, 2, or 3 coins at any stage of the game, depending on the number of coins left being equivalent to 2, 3, or 0 modulo 4.

Let n be the number of coins at a point in the game when it is player 1's turn to act, but player 1 has not yet taken the coins (such as at the start of the game). We can write n as $n = 4k + i$ for $i = 0, 1, 2, 3$.

First consider the following cases where $k = 0$, so $n = i$:

- If $i = 2$, player 1 will win by choosing 1 coin.

- If $i = 3$, player 1 will win by choosing 2 coins.
- If $i = 0$, player 1 has won because player 2 has already taken the last coin.

The next step is induction on k. Assume that it is known that if there are $4p+i$ coins left on the table when it is player 1's turn to act, with $0 \leq p \leq k$, and $i = 0, 2$, or 3, then player 1 wins. Suppose that $i = 0, 2$, or 3 and that there are $n = 4(k+1) + i$ coins on the table, so there are at least 4 coins, and it is player 1's turn to act. Player 1's strategy is to take $i - 1$ modulo 4 coins (so if $i = 0$, then player 1 takes 3 coins). Then, player 2 will take j coins, where j is equal to 1, 2, or 3. The remaining number of coins is then equivalent modulo 4 to

$$(4(k+1) + i) - (i - 1) - j \equiv 1 - j \pmod{4},$$

which is $1 - 1 \equiv 0$, $1 - 2 \equiv 3$, or $1 - 3 \equiv 2$ modulo 4. In particular,

- If $i = 0$ and $j = 2$ or 3, then there are $4(k+1) - 3 - j = 4(k-1) + (5-j)$ coins left.
- If $i = 0$ and $j = 1$, then there are $4(k+1) - 3 - 1 = 4k + 0$ coins left.
- If $i \neq 0$ and $j = 2$ or 3, then there are $4(k+1) + i - (i-1) - j = 4k + (5-j)$ coins left.
- If $i \neq 0$ and $j = 1$, then there are $4(k+1) + i - (i-1) - 1 = 4(k+1)$ coins left.

In the first three cases, the number of coins left is $4p + i'$, with $p = k - 1$ or k, and $i' = 0, 2$, or 3, so by the inductive hypothesis, player 1 will win. In the last case, the number of coins left is $4(k+1) + 0 \geq 4$, so the game is not over and the next moves by players 1 and 2 will fall into one of the first two cases where $i = 0$, and then the inductive hypothesis will apply.

S2010-4 If the rank of B is 0, then $B = 0$, so $A = I$ follows from $A + B = I$ and the required conditions hold. Otherwise, if the rank of B is p with $0 < p \leq n$, then by the Rank + Nullity Theorem for A, the dimension of the nullspace of A is p, and there is an ordered basis for the nullspace of A, $(\vec{a}_1, \ldots, \vec{a}_p)$. For each of the nullspace vectors, $B(\vec{a}_k) = (I - A)(\vec{a}_k) = \vec{a}_k - \vec{0} = \vec{a}_k$. So, $(\vec{a}_1, \ldots, \vec{a}_p)$ is an independent list of vectors in the p-dimensional image of B and must be a basis of the image of B. For any $\vec{x}$, $A(B(\vec{x})) = A(c_1\vec{a}_1 + \ldots + c_p\vec{a}_p) = \vec{0}$. From $AB = 0$, we can conclude $A(I - A) = 0 \implies A - A^2 = 0 \implies A = A^2$, and $(I - B)B = 0 \implies B - B^2 = 0 \implies B = B^2$. Finally, $BA = (I - A)A = A - A^2 = 0 = AB$.

S2010-5 The quotient is not defined at $x = 0$. The proof will proceed by noting the Taylor series expansions of the various functions and composing these appropriately. The Taylor series for the relevant functions are

$$\sin x = x - \frac{x^3}{3!} + \text{Terms of at least order 5},$$

$$\tan x = x + \frac{x^3}{3} + \text{Terms of at least order 5},$$

$$\arcsin x = x + \frac{x^3}{3!} + \text{Terms of at least order 5},$$

$$\arctan x = x - \frac{x^3}{3} + \text{Terms of at least order 5}.$$

After composition and some manipulation, we obtain the following limit:

$$\lim_{x \to 0} \frac{-x^3 + \text{Terms of at least order 5}}{x^3 + \text{Terms of at least order 5}}$$

As a consequence, the limit is -1.

S2010-6 The thing to observe is that we are looking for ordered pairs (x, y) in the first quadrant that satisfy the inequality $\frac{1}{2} < \frac{y}{x} < 3$ whose sum is odd and bounded above by 5. There are only three ordered pairs that satisfy these conditions: $(1, 2)$, $(2, 3)$, and $(3, 2)$. $U + D + R + L = 5$, where the letters represent the number of moves up (U), down (D), and so forth. Let the ordered quadruple (U, D, R, L) represent the set of directions. To get to $(2, 3)$, $U = 3$ and $R = 2$. The associated probability is $\frac{5!}{2!3!}(\frac{1}{16})^2(\frac{1}{4})^3$. Similarly, the probability associated with the point $(3, 2)$ is $\frac{5!}{3!2!}(\frac{1}{16})^3(\frac{1}{4})^2$. Both $(2, 0, 2, 1)$ and $(3, 1, 1, 0)$ will reach $(1, 2)$. The probability associated with the first quadruple is $\frac{5!}{2!2!1!}(\frac{1}{4})^2(\frac{1}{16})^2(\frac{9}{16})$. The probability associated with the second quadruple is $\frac{5!}{3!1!1!}(\frac{1}{4})^3(\frac{1}{8})(\frac{1}{16})$. The sum of these four probabilities is the solution.

S2010-7 The difficulty is in finding a way to show that for rational numbers close to a given irrational number x, the denominator gets large. We may also assume that $x \in [0, 1]$ and that x is irrational. Let $\epsilon > 0$. There is an $n \in \mathbb{N}$ such that $\frac{1}{n} < \epsilon$. For every $1 \leq i \leq n$, define

$$\delta_i = \min\left\{|x - 0|, \left|x - \frac{1}{i}\right|, \ldots, \left|x - \frac{i-1}{i}\right|, |x - 1|\right\}.$$

Define $\delta = \frac{1}{2}\min\{\delta_i\}$; it follows from the irrational property of x that $\delta > 0$. The claim is that if $|y - x| < \delta$, then $|f(y) - f(x)| < \epsilon$, so by definition, f is continuous at x. To prove the claim, consider $y \in (x - \delta, x + \delta)$.

Case 1. If $y \notin \mathbb{Q}$, then $|f(x) - f(y)| = 0$.

Case 2. If $y \in \mathbb{Q}$, with $y = \frac{p}{q}$ in the lowest terms and $q \leq n$, then $|x - \frac{p}{q}| < \delta < \delta_q \leq |x - \frac{p}{q}|$, a contradiction.

Case 3. The only remaining case is that $y \in \mathbb{Q}$, with $y = \frac{p}{q}$ in the lowest terms and $q > n$, so $|f(y) - f(x)| = |\frac{1}{q} - 0| < \frac{1}{n} < \epsilon$.

Chapter 34
2011 Solutions

Indiana Wesleyan University hosted an Indiana MAA meeting and ICMC event for the first time in 2011.

The 2011 problem set was prepared by Professor Chris Mitchell at American University.

The **2011 problem statements** begin on page 23 in Chap. 11.

S2011-1 No, there is no even number $n \geq 6$ such that it is possible to achieve any elimination order by choosing a corresponding elimination parameter. Consider, for any even number $n \geq 6$, the following order of elimination: $1, n, n-1, n-2, \ldots, 3$, so that child 2 is the winner. If the first child on the list is person 1, then the parameter i must be of the form nk, for some integer k; therefore the elimination parameter must be even. If the game continues until only players 2, 3, and 4 remain, then according to the specified order, child 4 leaves, and the count starts again with child 2. However, any even count starting with child 2 would end with child 3 and require child 2 to leave, contradicting the chosen elimination order.

S2011-2 If x represents the length of the cut, let $P(x)$, $A(x)$, $S(x)$, and $V(x)$ represent the perimeter of the base, the area of the base, the area of the sides, and the volume of the resulting box, as functions of x.

Drawing a line segment from a vertex of the interior triangle to the corresponding vertex of the exterior triangle creates two $30° - 60° - 90°$ triangles. Since the short leg has length x, we can conclude the hypotenuse has length $2x$ and the long leg has length $\sqrt{3}x$. It follows that each side of the interior triangle has length $1 - 2\sqrt{3}x$, $P(x) = 3(1 - 2\sqrt{3}x)$, and $S(x) = 3x(1 - 2\sqrt{3}x)$. The lengths x and $1 - 2\sqrt{3}x$ are nonnegative on the domain interval $[0, \frac{1}{2\sqrt{3}}]$.

Drawing an altitude from one vertex of the interior triangle to its opposite side yields another pair of $30° - 60° - 90°$ triangles. Using a similar method as before, we can conclude the altitude length (i.e., height) of the interior triangle is $\frac{\sqrt{3}}{2}(1 - 2\sqrt{3}x)$ and, thus, $A(x) = \frac{\sqrt{3}}{4}(1 - 2\sqrt{3}x)^2$. It immediately follows that $V(x) = \frac{\sqrt{3}}{4}x(1 - 2\sqrt{3}x)^2$.

The result is implied by maximizing $V(x)$ on $[0, \frac{1}{2\sqrt{3}}]$ using the tools of calculus. Note that $V'(x) = 9\sqrt{3}x^2 - 6x + \frac{\sqrt{3}}{4}$. Setting the derivative equal to 0 and solving for x (using the quadratic formula) yields $x = \frac{1}{2\sqrt{3}}$ or $x = \frac{1}{6\sqrt{3}}$. The former obviously produces a minimal volume because $V(\frac{1}{2\sqrt{3}}) = 0$; the latter value for x uniquely produces a maximal volume. The two lengths $x = \frac{1}{2\sqrt{3}}, x = \frac{1}{6\sqrt{3}}$, are also the only solutions of $A(x) = S(x)$.

Finally, we note that $A(\frac{1}{6\sqrt{3}}) = \frac{1}{3\sqrt{3}} = S(\frac{1}{6\sqrt{3}})$, which completes the problem.

S2011-3 For a particular number n, the function defined for $(x, y) \neq (0, 0)$ by

$$f(x, y) = \frac{x^{n+1}y}{x^{2(n+1)} + y^2}$$

(and then extended to the domain $\mathbb{R}^2$ by $f(0, 0) = 0$) satisfies

$$\lim_{x \to 0} f(x, g(x)) = 0$$

for all $g \in C^n$. Let $y = x^{n+1}$; then

$$\lim_{x \to 0} f(x, y) = \lim_{x \to 0} \frac{x^{2n+2}}{x^{2n+2} + x^{2n+2}} = \frac{1}{2} \neq 0.$$

Thus, in particular, it is not true that

$$\lim_{(x,y) \to (0,0)} f(x, y) = 0,$$

and in fact the limit does not exist.

Remark from the Editors The $n = 1$ case of this example appears in ICMC **P1986-6** [AFMC].

S2011-4 If n and b are relatively prime, then there is a minimal nonzero r for which $b^r \equiv 1 \pmod{n}$, or $b^r - 1 = kn$, for some k. It follows that

$$k = d_{r-1}b^{r-1} + \cdots d_1 b + d_0 \text{ for } 0 \leq d_i < b.$$

Putting the last two ideas together, we obtain

$$\frac{1}{n} = \frac{d_{r-1}b^{r-1} + \cdots d_1 b + d_0}{b^r} \cdot \frac{1}{1-b^{-r}}.$$

Since $\frac{1}{1-b^{-r}} = \sum_{i=0}^{\infty} (b^{-r})^i$, a cycle length for $\frac{1}{n}$ is r. To show that there is no lower cycle length, note that the equation (which would follow if m were another cycle length)

$$\frac{1}{n} = \frac{e_{m-1}b^{m-1} + \cdots e_1 b + e_0}{b^m} \cdot \frac{1}{1-b^{-m}}$$

implies that $m \geq r$ because of the minimality of r.

S2011-5 Imagine that the three pegs are occupied, from left to right, by the red stack and then the blue stack, and the rightmost peg is unoccupied. Label these three pegs P_L, P_M, and P_R (for left peg, middle peg, and right peg). In order to swap the positions of the two stacks, we will create a double stack containing $2(n-1)$ disks of alternating color (the pattern will be, from the top; red, blue, red, blue, and so on) over the largest blue disk on P_M. We will then be able to move: 1. the largest red disk to P_R; 2. the double stack over the largest red disk on P_R; 3. the largest blue disk onto P_L; 4. the double stack onto the largest blue disk on P_L; 5. the largest red disk onto P_M. Finally, we unstack the double stack. Let $D_M(k)$ denote the number of moves required to create a double stack of $2k$ disks on P_M. Let $A(k)$ denote the number of moves required to implement the entire algorithm. Since it will require $2(2^k - 1)$ moves to shift the double stack from one peg to another, we would have the following:

$$A(n) = 2D_M(n-1) + 3 + 4(2^{n-1} - 1).$$

Suppose that one has two stacks of k red and blue disks. Let $D_R(k)$ denote the number of moves required to shuffle these two stacks into a single double stack which sits atop P_R. In order to create a double stack of $2k$ disks over the P_M, one is required to create a double stack of $2(k-1)$ disks over P_R. Once this has been done, the second largest red disk would be moved atop the second largest blue disk on P_M, and then the double stack of $2(k-1)$ disks would be moved from P_R onto P_M. We would then have

$$D_M(k) = D_R(k-1) + 2(2^{k-2} - 1) + 1.$$

On the other hand, in order to create a double stack of $2k$ disks on P_R, we are required to create a double stack of $2(k-1)$ disks on P_M, move a red disk from P_L peg onto P_R, move the double stack of $2(k-1)$ disks onto the leftmost peg, move a blue disk from the center peg onto the rightmost peg, and then move the double stack onto the rightmost peg. So:

$$D_R(k) = D_M(k-1) + 2 + 4(2^{k-1} - 1).$$

Putting these two together, we would have the relation:

$$D_M(k) - D_M(k-2) = 3 + 2(2^{k-2} - 1) + 4(2^{k-2} - 1).$$

Since $D_M(0) = 0$ and $D_M(1) = 1$, this relation permits an efficient calculation of $D_M(k)$ for any k by examining the telescoping sum (for $i = 2$ or $i = 3$):

$$(D_M(k) - D_M(k-2)) + (D_M(k-2) - D_M(k-4)) + \cdots$$
$$\cdots + (D_M(i) - D_M(i-2))$$
$$= D_M(k) - D_M(i-2).$$

For even k write $k = 2j$. Then

$$D_M(2j) - D_M(0) = \sum_{i=1}^{j} (D_M(2i) - D_M(2i-2))$$
$$= \sum_{i=1}^{j} \left(3 \cdot 2^{2i-1} - 3\right)$$
$$= 3 \cdot \frac{4^{j+1} - 4}{2 \cdot 3} - 3j$$
$$= 2 \cdot 2^{2j} - 2 - 3j$$
$$= 2 \cdot 2^k - 2 - \frac{3}{2}k.$$

For odd k write $k = 2j + 1$. Then

$$D_M(2j+1) - D_M(1) = \sum_{i=1}^{j} (D_M(2i+1) - D_M(2i-1))$$
$$= \sum_{i=1}^{j} \left(3 \cdot 2^{2i} - 3\right)$$
$$= 3 \cdot \frac{4^{j+1} - 4}{3} - 3j$$
$$= 2^{2j+2} - 4 - 3j$$

$$= 2 \cdot 2^k - \frac{5}{2} - \frac{3}{2}k.$$

Thus $D_M(2j+1) = D_M(1) + 2 \cdot 2^k - \frac{5}{2} - \frac{3}{2}k = 2 \cdot 2^k - \frac{3}{2} - \frac{3}{2}k$.

Therefore $D_M(k) = \begin{cases} 2^{k+1} - 2 - \frac{3}{2}k & \text{if } k \text{ is even,} \\ 2^{k+1} - \frac{3}{2} - \frac{3}{2}k & \text{if } k \text{ is odd.} \end{cases}$

Alternatively this may be written as $D_M(k) = 2^{k+1} - \frac{3}{2}k - \frac{7+(-1)^k}{4}$. Substitute this into the formula for $A(n)$ to get

$$A(n) = 2D_M(n-1) + 3 + 4(2^{n-1} - 1)$$
$$= 2^{n+2} - 3n - \frac{3 - (-1)^n}{2}.$$

S2011-6 Assume that $x_n \neq f(x_n)$ for all $n \geq 1$. Otherwise, a fixed point clearly exists, and the limit converges to this fixed point. In particular, $x_1 \neq f(x_1)$. By the Mean Value Theorem, there is some x_1^* between x_1 and $f(x_1) = x_2$ so that

$$\frac{f(x_2) - f(x_1)}{x_2 - x_1} = f'(x_1^*)$$
$$\implies \frac{|f(x_2) - f(x_1)|}{|x_2 - x_1|} = |f'(x_1^*)| \leq \frac{1}{2}$$
$$\implies |x_3 - x_2| \leq \frac{|x_2 - x_1|}{2}.$$

More generally for $n > 1$, using $x_n \neq x_{n-1}$,

$$\frac{|f(x_n) - f(x_{n-1})|}{|x_n - x_{n-1}|} \leq \frac{1}{2}$$
$$\implies |x_{n+1} - x_n| \leq \frac{|x_n - x_{n-1}|}{2},$$

and by induction,

$$|x_{n+1} - x_n| \leq \frac{|x_2 - x_1|}{2^{n-1}} \tag{34.1}$$

for all $n \geq 1$. For $n > m + 1 \geq 2$, using the triangle inequality and (34.1):

$$|x_n - x_m| \leq |x_n - x_{n-1}| + |x_{n-1} - x_{n-2}| + \cdots + |x_{m+1} - x_m|$$
$$\leq |x_2 - x_1| \cdot \left(\frac{1}{2^{n-2}} + \frac{1}{2^{n-3}} + \cdots + \frac{1}{2^{m-1}}\right)$$

$$< \frac{|x_2 - x_1|}{2^{m-2}}.$$

To summarize, the inequality $|x_n - x_m| < \frac{|x_2 - x_1|}{2^{m-2}}$ holds for all $n \geq m \geq 1$. So, for any $\epsilon > 0$, choose $N \geq 1$ so that $\frac{|x_2 - x_1|}{2^{N-2}} < \epsilon$. Then for $n \geq m \geq N$,

$$|x_n - x_m| < \frac{|x_2 - x_1|}{2^{m-2}} \leq \frac{|x_2 - x_1|}{2^{N-2}} < \epsilon.$$

This implies that the sequence is Cauchy, therefore convergent. Since the limit exists and f is continuous, it must be that

$$f(\lim_{n \to \infty} x_n) = \lim_{n \to \infty} f(x_n) = \lim_{n \to \infty} x_n.$$

S2011-7 The following function $f : \mathbb{R} \to \mathbb{R} \setminus \{0\}$ is a one-to-one correspondence:

$$f(x) = \begin{cases} x & \text{if } x \text{ is not a whole number,} \\ x+1 & \text{if } x \text{ is a whole number.} \end{cases}$$

S2011-8 The matrix in question can be thought of as the evaluation of a polynomial with complex coefficients at the following matrix X:

$$\begin{bmatrix} 0 & 0 & 0 & \cdots & 1 \\ 1 & 0 & 0 & \cdots & 0 \\ 0 & 1 & 0 & \cdots & 0 \\ \vdots & \vdots & \vdots & \ddots & \vdots \\ 0 & 0 & \cdots & 1 & 0 \end{bmatrix}.$$

To be more precise, if A is the matrix described in the problem and I is the identity matrix, then

$$A = I + nX + (n-1)X^2 + \cdots + 2X^{n-1}.$$

It follows that if v is an eigenvector for X, then v is an eigenvector for A. From $\det(X - \lambda I) = \pm(1 - \lambda^n)$, we see that X has n distinct complex eigenvalues (the nth roots of unity), and therefore there exists a list of n eigenvectors for X which is linearly independent (over $\mathbb{C}$). Such a list of eigenvectors for X is a complete list of eigenvectors for A. Any eigenvector v for X is a (nonzero) complex scalar multiple of a vector of the form

$$\begin{bmatrix} 1 \\ \omega_k \\ \omega_k^2 \\ \vdots \\ \omega_k^{n-1} \end{bmatrix},$$

where ω_k is one of the nth roots of unity, $\{\omega_0, \omega_1, \ldots, \omega_{n-1}\}$ (including $\omega_0 = 1$).

Chapter 35
2012 Solutions

The 47th competition was held at Ball State University.

The 2012 problem set was prepared by Professor Ahmed Mohammed at the host institution.

The **2012 problem statements** begin on page 27 in Chap. 12.

S2012-1 Obviously, the statement is true for $n = 1$. So we assume that $n \geq 2$. By the Binomial Theorem we note that

$$(n+1)^n - 1 = \left(\sum_{j=0}^{n} \binom{n}{j} n^j\right) - 1 = \sum_{j=1}^{n} \binom{n}{j} n^j$$

$$= \binom{n}{1} n + \sum_{j=2}^{n} \binom{n}{j} n^j = n^2 + \sum_{j=2}^{n} \binom{n}{j} n^j.$$

On noting that $\binom{n}{j}$ is a positive integer, we see that

$$\sum_{j=1}^{n} \binom{n}{j} n^j$$

is divisible by n^2. Therefore, indeed $(n+1)^n - 1$ is divisible by n^2.

S2012-2 First we observe that the number of zeros at the end of 213! is same as the number of times the number 10 occurs as a factor of 213!. More precisely, since $10 = 2 \times 5$, and since there are more factors of 2 than 5 in the unique prime factorization of the product 213!, we observe that there are as many ending zeros in the product 213! as there are factors of 5. The number of positive multiples of 5 less or equal to 213 is

$$\left[\frac{213}{5}\right] = 42, \tag{35.1}$$

where $[x]$ denotes the greatest integer less than or equal to x. Among these multiples, we list those that contain two or more factors of 5 as follows:

$$25 = 5^2, \quad 50 = 2 \cdot 5^2, \quad 75 = 3 \cdot 5^2, \quad 100 = 4 \cdot 5^2$$
$$125 = 5^3, \quad 150 = 6 \cdot 5^2, \quad 175 = 7 \cdot 5^2, \quad 200 = 8 \cdot 5^2.$$

Therefore there are exactly 9 additional occurrences of the digit 5 that have not been counted in (35.1). The conclusion is that there are a total of $42 + 9 = 51$ occurrences of the digit 5 in the product 213!, and so 51 zeros at the end of 213!.

Remark from the Editors For a general approach to finding the number of times a prime p occurs as a factor of a factorial $m!$, see the solution of Problem #2 of the 1925 Eötvös Competition in Hungary [K$_2$].

S2012-3

(a) Suppose r/s is a rational root so that r and s are integers with $s \neq 0$. Clearly $r \neq 0$. We suppose that $\gcd(r, s) = 1$, so that the fraction r/s is in lowest terms, and in particular, r and s cannot both be even. We proceed to show that all other parities lead to a contradiction.

By assumption we have

$$0 = p\left(\frac{r}{s}\right) = a_n \left(\frac{r}{s}\right)^n + \cdots + a_2 \left(\frac{r}{s}\right)^2 + a_1 \left(\frac{r}{s}\right) + a_0.$$

Clearing fractions, we see that

$$a_n r^n + a_{n-1} s r^{n-1} + \cdots a_2 s^{n-2} r^2 + a_1 s^{n-1} r + a_0 s^n = 0. \tag{35.2}$$

Case 1. Suppose either r or s is even and the other is odd. This would imply that every LHS term of (35.2) except one is even, with the remaining term odd, contradicting RHS being even.

Case 2. Suppose now both s and r are odd. Add the odd integer $a_n + \cdots + a_2$ to both sides of (35.2) and get

$$a_n(1+r^n) + a_{n-1}(1+sr^{n-1}) + \cdots + a_2(1+s^{n-2}r^2) + a_1s^{n-1}r$$
$$+ a_0 s^n$$
$$= a_n + \cdots + a_2. \tag{35.3}$$

Then $1 + s^i r^j$ is an even integer for any nonnegative integers i, j. Therefore $a_n(1+r^n) + a_{n-1}(1+sr^{n-1}) + \cdots + a_2(1+s^{n-2}r^2)$ is even. Since a_0 and a_1 are both odd, we note that each of $a_1 s^{n-1} r$ and $a_0 s^n$ is odd, and therefore their sum is even. Therefore we see that LHS of (35.3) is even. But we recall that RHS is odd. Thus, once again, we conclude that r and s cannot be both odd.

(b) Each of the following polynomials has a rational root:

$$p_1(x) = 1x^2 + 1x - 2 \implies p_1(1) = 0$$
$$p_2(x) = 1x^2 - 2x + 1 \implies p_2(1) = 0$$
$$p_3(x) = 2x^3 + 1x^2 + 1x - 1 \implies p_3(\frac{1}{2}) = 0$$
$$p_4(x) = 1x^3 + 1x^2 - 1x - 1 \implies p_4(1) = 0.$$

These examples show that none of the conditions in the problem may be omitted.

Remark from the Editors For the polynomial $p(x) = a_n x^n + \ldots + a_0$ with a rational root r/s in lowest terms, the Rational Root Theorem (see **S2001-4A** in Chap. 24) states that r is a divisor of a_0 and s is a divisor of a_n. So if a_n and a_0 are both odd, then r and s are also both odd; the above Case 1. argument is a special case of this idea.

S2012-4A First let us dispose of the trivial case when $\beta = 0$. In this case $A = \alpha I_n$, which is obviously diagonalizable, and $\det A = \alpha^n$. So, henceforth we suppose that $\beta \neq 0$. Note that $A = \vec{w}\vec{v}^T - \gamma I_n$, where

$$\vec{w} = \begin{bmatrix} 1 \\ 1 \\ \vdots \\ 1 \end{bmatrix}, \quad \vec{v} = \begin{bmatrix} \beta \\ \beta \\ \vdots \\ \beta \end{bmatrix}, \quad \text{and } \gamma = \beta - \alpha.$$

Now, a nonzero vector $\vec{x}$ in $\mathbb{R}^n$ is an eigenvector of A if and only if $A\vec{x} = \lambda \vec{x}$ for some $\lambda \in \mathbb{R}$. That is,

$$\vec{w}\vec{v}^T \vec{x} - \gamma \vec{x} = \lambda x.$$

We rewrite this as

$$(\vec{v}^T\vec{x})\vec{w} = (\lambda + \gamma)\vec{x}.$$

If $c = \vec{v}^T\vec{x} \neq 0$, then note that $\lambda + \gamma \neq 0$, and therefore

$$\vec{x} = c(\lambda + \gamma)^{-1}\vec{w},$$

showing that $\vec{x}$ is a multiple of $\vec{w}$. If $\vec{v}^T\vec{x} = 0$, then $\vec{x}$ is orthogonal to $\vec{v}$ and hence to $\vec{w}$ (recall that $\beta \neq 0$). Thus, any eigenvector of A is either a multiple of $\vec{w}$ or orthogonal to $\vec{w}$. Therefore we see that $\vec{w}$ is an eigenvector of A with

$$A\vec{w} = \vec{w}\vec{v}^T\vec{w} - \gamma\vec{w} = (n\beta - \gamma)\vec{w} = ((n-1)\beta + \alpha)\vec{w}.$$

That is, $\lambda = (n-1)\beta + \alpha$ is an eigenvalue of A with corresponding eigenspace of dimension 1. On the other hand, any other eigenvector of A must be orthogonal to $\vec{w}$ with corresponding eigenvalue $\lambda = -\gamma = \alpha - \beta$. The eigenspace of A corresponding to $\lambda = \alpha - \beta$ is the orthogonal complement of the eigenspace of A corresponding to the eigenvalue $\lambda = (n-1)\beta + \alpha$, namely the orthogonal complement of the line parallel to $\vec{w}$. Thus the eigenspace of A corresponding to the eigenvalue $\lambda = \alpha - \beta$ has dimension $n-1$. This shows that A is diagonalizable. The determinant of A is the product of its eigenvalues

$$\det A = (\alpha - \beta)^{n-1}((n-1)\beta + \alpha).$$

S2012-4B As in the previous solution, let $\vec{w} = [1, 1, \ldots, 1]^T$, so that $A = (\alpha - \beta)I_n + \beta\vec{w}\vec{w}^T$. Let $\vec{e}_1 = [1, 0, \ldots, 0]^T$ be the first standard basis column vector, and let P be the invertible matrix where the first column of I_n is replaced by $\vec{w}$:

$$P = \begin{bmatrix} 1 & & & \\ 1 & 1 & & \\ \vdots & & \ddots & \\ 1 & & & 1 \end{bmatrix},$$

so that $P\vec{e}_1 = \vec{w}$ and $P^{-1}\vec{w} = \vec{e}_1$. Then A is similar to an upper triangular matrix:

$$P^{-1}AP = (\alpha - \beta)I_n + \beta P^{-1}\vec{w}\vec{w}^T P$$
$$= (\alpha - \beta)I_n + \beta\vec{e}_1[n, 1, \ldots, 1]$$
$$= \begin{bmatrix} \alpha + (n-1)\beta & \beta & \cdots & \beta \\ & \alpha - \beta & & \\ & & \ddots & \\ & & & \alpha - \beta \end{bmatrix}.$$

By inspection, $\det(P^{-1}AP) = (\alpha - \beta)^{n-1}(\alpha + (n-1)\beta)$, $\vec{e}_1$ is an eigenvector of $P^{-1}AP$ with eigenvalue $\alpha + (n-1)\beta$, and there are $n-1$ eigenvectors of the form $[1, 0, \ldots, 0, -n, 0, \ldots, 0]^T$ with eigenvalue $\alpha - \beta$. So, $P^{-1}AP$ has n independent eigenvectors and is diagonalizable, and it follows that A is also diagonalizable with the same determinant.

S2012-5 Let $N = \{e, a\}$ be a normal subgroup of G of order 2, where e is the identity of the group. It follows that a has order 2. Since N is a normal subgroup of G, by definition, we see that $g^{-1}ag \in G$ for any $g \in G$. Since $a \neq e$, we must have $g^{-1}ag = a$. That is, $ag = ga$. Hence we have shown that $ag = ga$ for all $g \in G$. Since $ag = ga$, one can easily show by the principle of mathematical induction that $(ag)^n = a^n g^n$ for any $g \in G$, and any nonnegative integer n. The quotient group G/N is of order 13, and since any group of prime order is cyclic, this quotient group is cyclic. Let $b \in G\setminus N$. Then the coset bN has order 13. In particular the order of b cannot be 2. Since its order has to divide 26, the order of b must be either 13 or 26. If the order is 26, then G is cyclic with generator b. If the order is 13, then since $ab = ba$, we must have $(ab)^{13} = a^{13}b^{13} = a$, and therefore $(ab)^{26} = a^2 = e$, and hence ab has order 26. Therefore ab generates G, and therefore G is cyclic.

S2012-6 Suppose contrary to what is asserted, the indicated sum is a rational number. Then there are positive integers a and b such that

$$\sum_{n=0}^{\infty} \frac{1}{(n!)^k} = \frac{a}{b}.$$

We multiply both sides by $((b+1)!)^k$ and rewrite the sum as follows:

$$((b+1)!)^k \frac{a}{b} = \sum_{n=0}^{\infty} \frac{((b+1)!)^k}{(n!)^k}$$

$$= \sum_{n=0}^{b+1} \frac{((b+1)!)^k}{(n!)^k} + \sum_{n=b+2}^{\infty} \frac{((b+1)!)^k}{(n!)^k}.$$

Thus

$$\sum_{n=b+2}^{\infty} \frac{((b+1)!)^k}{(n!)^k} = ((b+1)!)^k \frac{a}{b} - \sum_{n=0}^{b+1} \frac{((b+1)!)^k}{(n!)^k}. \tag{35.4}$$

It is clear that the right-hand side is a positive integer. We now proceed to show that the left-hand side is not an integer, thereby getting the desired contradiction. For $n \geq b+2$, we write $n = (b+2) + j$ for $j \geq 0$ and hence

$$\frac{((b+1)!)^k}{(n!)^k} = \left(\frac{(b+1)!}{((b+2)+j)!}\right)^k$$

$$= \left(\frac{1}{(b+2+j)\cdots(b+2)}\right)^k$$

$$\leq \left(\frac{1}{(b+2)^{j+1}}\right)^k.$$

As a consequence we see that

$$\sum_{n=b+2}^{\infty} \frac{((b+1)!)^k}{(n!)^k} = \sum_{j=0}^{\infty} \frac{((b+1)!)^k}{(((b+2)+j)!)^k}$$

$$\leq \sum_{j=0}^{\infty} \left(\frac{1}{(b+2)^k}\right)^{j+1}$$

$$= \frac{1}{(b+2)^k - 1}.$$

Note that $\dfrac{1}{(b+2)^k - 1} < 1$, showing that the left-hand side sum in (35.4) cannot be an integer.

S2012-7 Note that

$$\int_0^1 nx(1-x^2)^n\, dx = -\frac{n}{2(n+1)}(1-x^2)^{n+1}\Big|_0^1 = \frac{n}{2(n+1)}. \tag{35.5}$$

As a result of this, the claim will follow once we show that

$$\lim_{n\to\infty} \int_0^1 nx(1-x^2)^n[f(x) - f(0)]\, dx = 0.$$

To this end, let $\epsilon > 0$ be given. Since f is continuous at $x = 0$, there is, corresponding to the given ϵ (but not depending on n), a number $\delta \in (0, 1)$ such that if $0 \leq x < \delta$, then $|f(x) - f(0)| < \epsilon$. First let us note that

$$\int_0^1 nx(1-x^2)^n |f(x) - f(0)|\, dx = \int_0^\delta nx(1-x^2)^n |f(x) - f(0)|\, dx$$

$$+ \int_\delta^1 nx(1-x^2)^n |f(x) - f(0)|\, dx$$

$$= I_n + II_n. \tag{35.6}$$

We estimate each of the summands I_n and II_n. For I_n, we use the continuity of f at 0. Thus for any n,

$$I_n = \int_0^\delta nx(1-x^2)^n |f(x)-f(0)|\,dx \leq \epsilon \int_0^1 nx(1-x^2)^n\,dx$$

$$= \frac{\epsilon n}{2(n+1)}$$

$$< \epsilon/2.$$

Next we estimate II_n. For this, we use the boundedness of f on $[0,1]$: There is some $M > 0$ so that $|f(x)| \leq M$ for all $0 \leq x \leq 1$. For any $n > 0$, the function $g(x) = n(1-x^2)^n$ is decreasing on $[0,1]$, so on the interval $[\delta, 1]$,

$$0 \leq g(x) \leq n(1-\delta^2)^n.$$

We estimate:

$$II_n = \int_\delta^1 nx(1-x^2)^n |f(x)-f(0)|\,dx$$

$$\leq 2M \int_\delta^1 nx(1-x^2)^n\,dx$$

$$\leq 2M \int_\delta^1 nx(1-\delta^2)^n\,dx$$

$$< Mn(1-\delta^2)^n.$$

On noting that $\lim_{n\to\infty} n(1-\delta^2)^n = 0$, there is some N depending on δ and ϵ so that if $n > N$ then $II_n < \epsilon/2$. Therefore, from (35.6) and the above estimates we find that for sufficiently large n,

$$\left|\int_0^1 nx(1-x^2)^n (f(x)-f(0))\,dx\right| \leq \int_0^1 nx(1-x^2)^n |f(x)-f(0)|\,dx$$

$$= I_n + II_n \leq \epsilon.$$

Therefore, we have

$$\lim_{n\to\infty} \int_0^1 nx(1-x^2)^n (f(x)-f(0))\,dx = 0 \tag{35.7}$$

as claimed. Finally, adding $\lim_{n\to\infty} \int_0^1 nx(1-x^2)^n f(0)\,dx$ to both sides of (35.7), and using (35.5), we find that

$$\lim_{n\to\infty}\int_0^1 nx(1-x^2)^n f(x)\,dx = \lim_{n\to\infty}\int_0^1 nx(1-x^2)^n f(0)\,dx = \frac{1}{2}f(0).$$

S2012-8 First we rewrite the given double integral as

$$\iint_{\mathcal{R}} f(x,y)\,dx\,dy = \iint_{\mathcal{D}_1} f(x,y)\,dx\,dy - \iint_{\mathcal{D}_2} f(x,y)\,dx\,dy, \qquad (35.8)$$

where $\mathcal{D}_1$ and $\mathcal{D}_2$ are the closed disks

$\mathcal{D}_1 = \{(x,y) : x^2 + (y+1)^2 \le 9\}$ and $\mathcal{D}_2 = \{(x,y) : x^2 + (y-1)^2 \le 1\}$.

The following property of a function $f(x,y)$ that is harmonic in an open set containing a (closed, positive radius) disk $\mathcal{D}$ centered at (x_0, y_0) can be interpreted as a *Mean Value* property:

$$\frac{1}{\text{area}(\mathcal{D})}\iint_{\mathcal{D}} f(x,y)\,dx\,dy = f(x_0, y_0). \qquad (35.9)$$

We now apply formula (35.9) to (35.8) to get

$$\iint_{\mathcal{R}} f(x,y)\,dx\,dy = \iint_{\mathcal{D}_1} f(x,y)\,dx\,dy - \iint_{\mathcal{D}_2} f(x,y)\,dx\,dy$$
$$= \text{area}(\mathcal{D}_1) f(0,-1) - \text{area}(\mathcal{D}_2) f(0,1)$$
$$= 9\pi f(0,-1) - \pi f(0,1).$$

To prove the Mean Value Property (35.9), let r be the radius of $\mathcal{D}$, and consider any disk $\mathcal{E}$ of radius $0 < \rho \le r$ centered at (x_0, y_0). We give the curve $\partial\mathcal{E}$ a counterclockwise orientation and apply Green's Theorem to find

$$0 = \iint_{\mathcal{E}} [f_{xx}(x,y) + f_{yy}(x,y)]\,dx\,dy$$
$$= \iint_{\mathcal{E}} [(f_x)_x(x,y) - (-f_y)_y(x,y)]\,dx\,dy \qquad (35.10)$$
$$= \int_{\partial\mathcal{E}} \left[-f_y(x,y)\,dx + f_x(x,y)\,dy\right] \qquad (35.11)$$
$$= \int_0^{2\pi} \big(\rho f_y(x_0 + \rho\cos\theta, y_0 + \rho\sin\theta)\sin\theta$$
$$\quad + \rho f_x(x_0 + \rho\cos\theta, y_0 + \rho\sin\theta)\cos\theta\big)\,d\theta.$$

In the last integral we used the following parametrization for the counterclockwise oriented circle $\partial\mathcal{E}$.

$$x = x_0 + \rho \cos\theta, \quad y = y_0 + \rho \sin\theta, \quad 0 \le \theta \le 2\pi.$$

On dividing both sides of the last equation by ρ, and observing that the multivariable chain rule applies, we see that for any $0 < \rho < r$,

$$\begin{aligned} 0 &= \int_0^{2\pi} (f_y(x_0+\rho\cos\theta, y_0+\rho\sin\theta)\sin\theta + f_x(x_0+\rho\cos\theta, y_0+\rho\sin\theta)\cos\theta)\, d\theta \\ &= \int_0^{2\pi} \frac{d}{d\rho}(f(x_0+\rho\cos\theta, y_0+\rho\sin\theta))\, d\theta. \end{aligned} \qquad (35.12)$$

Integrating in polar coordinates, and using the Fundamental Theorem of Calculus, an interchange of integrals of continuous functions, and (35.12), we obtain

$$\begin{aligned} &\iint_{\mathcal{D}} f(x,y)\, dx\, dy \\ &= \int_0^{2\pi}\int_0^r f(x_0+R\cos\theta, y_0+R\sin\theta) R\, dR\, d\theta \\ &= \int_0^{2\pi}\int_0^r \left[f(x_0, y_0) + \int_{\rho=0}^R \frac{d}{d\rho}(f(x_0+\rho\cos\theta, y_0+\rho\sin\theta))\, d\rho \right] R\, dR\, d\theta \\ &= \int_0^{2\pi}\int_0^r f(x_0, y_0) R\, dR\, d\theta \\ &\quad + \int_0^r \int_{\rho=0}^R \int_0^{2\pi} \frac{d}{d\rho}(f(x_0+\rho\cos\theta, y_0+\rho\sin\theta))\, R\, d\theta\, d\rho\, dR \\ &= \text{area}(\mathcal{D}) \cdot f(x_0, y_0) + 0. \end{aligned}$$

This proves the claim (35.9).

Remark from the Editors It is enough to assume f is harmonic on an open set containing $\mathcal{D}_1$, but not enough to assume only that f is harmonic on some open set containing $\mathcal{R}$. The problem statement's hypothesis that f is twice continuously differentiable is used only to justify the application of Green's Theorem in the step equating line (35.10) with line (35.11). There are statements of Green's Theorem with weaker hypotheses (see [CZ]) that could be applied instead; for example, the step from (35.10) to (35.11) still follows from assuming only that $f_{xx} + f_{yy}$ is continuous and (f_x, f_y) is differentiable as a function $\mathbb{R}^2 \to \mathbb{R}^2$.

Chapter 36
2013 Solutions

The 48th competition was held at Indiana University East for the first time.

The 2013 problem set was prepared by a faculty committee at the host institution.

The **2013 problem statements** begin on page 29 in Chap. 13.

The 2013 competition's first place team from Valparaiso. Left to Right: Michael Stuck, Julia Yuan, and Timothy Goodrich

S2013-1 We show that the sequence is monotone increasing and bounded above. It then follows from the Monotone Convergence Theorem that the sequence converges. We show that for every $n \in \mathbb{N}$, we have $s_n < 2$ and $s_{n+1} \geq s_n$. This

is done by induction. For $n = 1$, we have $s_1 = 1 < 2$ and $s_2 = \sqrt{2} > 1 = s_1$. Assume that for some $n \in \mathbb{N}$, we have $s_n < 2$ and $s_{n+1} > s_n$. Then $s_{n+1} = \sqrt{1 + s_n} < \sqrt{1+2} < 2$. Also, $s_{n+2} = \sqrt{1 + s_{n+1}} > \sqrt{1 + s_n} = s_{n+1}$. By the principle of Mathematical Induction, we obtain that for every $n \in \mathbb{N}$, we have $s_n < 2$ and $s_{n+1} < s_n$. Because the sequence is bounded above and increasing, it converges to some positive real number. Let $s = \lim_{n \to \infty} s_n$. We observe that
$$s = \lim_{n \to \infty} s_n = \lim_{n \to \infty} s_{n+1} = \lim_{n \to \infty} \sqrt{1 + s_n} = \sqrt{1 + s}.$$ Solving the equation $s = \sqrt{1 + s}$ for $s > 0$ yields $s = \dfrac{1 + \sqrt{5}}{2}$.

S2013-2

(a) We let $C_0 \in \mathcal{C}$ be an arbitrary element of the collection. Because C_0 is compact, it is closed and bounded, by the Heine–Borel Theorem. Also, $K = \bigcap_{C \in \mathcal{C}} C \subseteq C_0$, so K is bounded. Each member C of the family $\mathcal{C}$ is compact and therefore closed. Therefore, the intersection $K = \bigcap_{C \in \mathcal{C}} C$ is closed. Because K is closed and bounded, it is compact by the Heine–Borel Theorem. □

(b) Let $C_n = [-n, n]$ for all $n \in \mathbb{N}$. Each set C_n is closed and bounded and is therefore compact. Let $\mathcal{C} = \{C_n : n \in \mathbb{N}\}$. Then $\bigcup_{C \in \mathcal{C}} C = \mathbb{R}$, which is not compact (it is not bounded).

S2013-3

(a) We distinguish two cases.

- If $m > n$, there are no one-to-one functions from A to B by the pigeonhole principle.
- If $m \leq n$, then the principle of multiplication shows there are $_nP_m$ or $P(n, m) = \dfrac{n!}{(n-m)!} = n \cdot (n-1) \cdots (n - m + 1)$ one-to-one functions from A to B. Equivalently, if one enumerates the elements in A as $(a_i)_{i=1}^m$ and the elements of B as $(b_j)_{j=1}^n$, there are clearly n available images for domain element a_1; without loss of generality, say this element is b_1. If one is to maintain the one-to-one property, there are now $n - 1$ available images for a_2; and without loss of generality, say the image is b_2. This pattern continues, yielding $n - 2$ possible images for a_3, $n - 3$ images for a_4, and so on. This also yields the answer $n \cdot (n - 1) \cdots (n - m + 1)$.

(b) If m is not equal to n, there is no one-to-one and onto function from A to B. If m is equal to n, then by the principle of multiplication there are $n!$ one-to-one and onto functions from A to B.

S2013-4 An element $a \in \mathbb{Z}_{pq}$ is a generator of $\mathbb{Z}_{pq}$ if and only if a and pq are relatively prime. Because p and q are primes, the elements of $\mathbb{Z}_{pq}$ that are not relatively prime to pq are the multiples of p and the multiples of q. The multiples of p are $p, 2p, \ldots, (q-1)p$ (i.e., $q-1$ multiples). Using a similar argument, we see that there are $p-1$ multiples of q. Also, 0 is not a generator. Therefore, there are $(q-1) + (p-1) + 1 = p + q - 1$ elements of $\mathbb{Z}_{pq}$ that are not generators, leaving $pq - p - q + 1$ elements that are generators.

S2013-5 Because the index of H in G is 2, there are exactly two left cosets of H in G, and there are exactly two right cosets. The left cosets are H itself and a coset of the form aH, for some $a \in G$. Likewise, the right cosets are H and a set of the form Hb for some $b \in G$. Observe that $H \cap aH = \emptyset$ and $H \cup aH = G$; likewise $H \cap Hb = \emptyset$ and $H \cup Hb = G$. It follows that $aH = Hb$, which means that the left and the right cosets of H coincide, making H a normal subgroup of G.

S2013-6A

(a)

n	1	2	3	4	5	6	7
f_n	1	1	2	3	5	8	13

(b) For $n = 1$, we observe $f_{2(1)+1} = f_3 = 2$ and $f_{1+1}^2 + f_1^2 = (1)^2 + (1)^2 = 2$.
For $n = 2$, we observe $f_{2(2)+1} = f_5 = 5$ and $f_{2+1}^2 + f_2^2 = (2)^2 + (1)^2 = 5$.
For $n = 3$, we observe $f_{2(3)+1} = f_7 = 13$ and $f_{3+1}^2 + f_3^2 = (3)^2 + (2)^2 = 13$.

(c) For $n = 1$ and $n = 2$, the formula has been verified in part (b). Therefore, the basis steps hold for mathematical induction. Now assume, for the strong form of mathematical induction, the identity holds for all values of n up to $n = k - 1$. Then

$$f_{2k-3} = f_{k-1}^2 + f_{k-2}^2$$

and

$$f_{2k-1} = f_k^2 + f_{k-1}^2.$$

Now we need to verify that the identity holds for $n = k$. In order to do this, we calculate f_{2k+1}.

$$\begin{aligned} f_{2k+1} &= f_{2k} + f_{2k-1} \\ &= f_{2k-1} + f_{2k-2} + f_{2k-1} \\ &= 2f_{2k-1} + (f_{2k-1} - f_{2k-3}) \\ &= 3f_{2k-1} - f_{2k-3}. \end{aligned}$$

Substituting the induction hypothesis, we can write the last expression as

$$\begin{aligned}
f_{2k+1} &= 3\left(f_k^2 + f_{k-1}^2\right) - f_{k-1}^2 - f_{k-2}^2 \\
&= 3f_k^2 + 2f_{k-1}^2 - (f_k - f_{k-1})^2 \\
&= 2f_k^2 + f_{k-1}^2 + 2f_k f_{k-1} \\
&= 2f_k^2 + (f_{k+1} - f_k)^2 + 2f_k (f_{k+1} - f_k) \\
&= 2f_k^2 + (f_{k+1} - f_k)(f_{k+1} - f_k + 2f_k) \\
&= 2f_k^2 + (f_{k+1} - f_k)(f_{k+1} + f_k) \\
&= f_{k+1}^2 + f_k^2.
\end{aligned}$$

This completes the induction step.

S2013-6B An alternate approach to (c) is to recall (from **P2003-6** in Chap. 3) that if Q is the matrix $\begin{bmatrix} 1 & 1 \\ 1 & 0 \end{bmatrix}$, then $Q^n = \begin{bmatrix} f_{n+1} & f_n \\ f_n & f_{n-1} \end{bmatrix}$. From

$$\begin{aligned}
Q^{2n} &= \begin{bmatrix} f_{2n+1} & f_{2n} \\ f_{2n} & f_{2n-1} \end{bmatrix} \\
= Q^n \cdot Q^n &= \begin{bmatrix} f_{n+1} & f_n \\ f_n & f_{n-1} \end{bmatrix} \cdot \begin{bmatrix} f_{n+1} & f_n \\ f_n & f_{n-1} \end{bmatrix} \\
&= \begin{bmatrix} f_{n+1}^2 + f_n^2 & f_{n+1}f_n + f_n f_{n-1} \\ f_{n+1}f_n + f_n f_{n-1} & f_n^2 + f_{n-1}^2 \end{bmatrix},
\end{aligned}$$

comparing the upper left entries establishes the claim.

S2013-7 Since all these numbers are positive, it is sufficient to prove

$$\frac{4mM}{(m+M)^2} \le \frac{4ab}{(a+b)^2}.$$

This is equivalent to proving

$$\frac{4\frac{m}{M}}{\left(1 + \frac{m}{M}\right)^2} \le \frac{4\frac{a}{b}}{\left(1 + \frac{a}{b}\right)^2}.$$

Consider the function $f(x) = \dfrac{4x}{(1+x)^2}$. This function is increasing on $[0, 1]$. To see this, we note that $\dfrac{d}{dx}\left(\dfrac{4x}{(1+x)^2}\right) = 4\dfrac{1-x}{(x+1)^3}$ exists for $x > -1$ and is positive for $0 < x < 1$.

From $0 < m \leq a$ and $M > 0$, we obtain $0 < \frac{m}{M} \leq \frac{a}{M}$. Because $b \leq M$, we also have $\frac{a}{M} \leq \frac{a}{b}$. Finally, $a \leq b$, so $\frac{a}{b} \leq 1$. Therefore, $0 < \frac{m}{M} \leq \frac{a}{b} \leq 1$.

Hence we have

$$f\left(\frac{m}{M}\right) = \frac{4\frac{m}{M}}{\left(1+\frac{m}{M}\right)^2} \leq f\left(\frac{a}{b}\right) = \frac{4\frac{a}{b}}{\left(1+\frac{a}{b}\right)^2}.$$

2013-8 We let F_5 denote the number of pentagons and F_6 denote the number of hexagons. We consider the soccer ball to be a convex polyhedron, with $F = F_5 + F_6$ faces, E edges, and V vertices. By Euler's Formula $V - E + F = 2$.

Each vertex is of valence 3. We may think of placing an observer on each vertex, and let the observers report the number of faces they see. Each observer reports seeing three faces. Each face is observed by as many observers as there are corners on the face, so we obtain $3V = 5F_5 + 6F_6$.

Now place an observer in every hexagon, and let them report the number of pentagons that border their hexagon. There are F_6 observers, each reporting three pentagons, for a total of $3F_6$ reports. Each pentagon is bordered by five hexagons, so each pentagon is reported by five different observers, and $3F_6 = 5F_5$.

Now place an observer into each of the faces, and let them report the number of edges they see. Each edge will be observed by two observers, so $2E = 5F_5 + 6F_6$.

Beginning with the equation $V - E + F = 2$ (and multiplying by 6), we obtain $6V - 6E + 6F = 12$. Substituting $3V = 5F_5 + 6F_6$ and $2E = 5F_5 + 6F_6$ and $F = F_5 + F_6$, we obtain $2(5F_5 + 6F_6) - 3(5F_5 + 6F_6) + 6(F_5 + F_6) = 12$, or $F_5 = 12$.

From the equation $3F_6 = 5F_5$, we get $F_6 = 20$. Therefore, the soccer ball has 12 black pentagons and 20 white hexagons.

Remark from the Editors Note that the $3F_6 = 5F_5$ equation was not used until the last paragraph, so more generally, any convex polyhedron with pentagon and hexagon faces and valence three vertices must have exactly 12 pentagons. In the 2013 competition, one team's submitted answer was a well-drawn picture of a semitransparent soccer ball exhibiting 12 pentagons and 20 hexagons. The grader assigned partial credit for showing existence but not uniqueness.

Chapter 37
2014 Solutions

The 49th competition was held at Indiana University—Purdue University Fort Wayne, coinciding with the celebration of the 50th anniversary of the founding of that campus.

The 2014 problem set was prepared by Professors Dan Coroian and Marc Lipman at the host institution.

The **2014 problem statements** begin on page 31 in Chap. 14.

A tied-for-first place team from Taylor University, with the IPFW Mastodon sculpture: L to R: Ethan Gegner, Josh Keirs, Claire Spychalla

S2014-1 The expression (14.1) simplifies to

$$f(x) = \frac{a^{3/2}x^{1/3}(x^{1/6} - a^{1/6})}{a^{1/4}(a^{3/4} - x^{3/4})} \qquad (37.1)$$

and is continuous on $[0, a) \cup (a, \infty)$, with the $x \to 0^+$ limit equal to 0. For the $x \to a$ limit, the $\frac{0}{0}$ form of L'Hôpital's Rule applies.

$$\lim_{x \to a} \frac{\sqrt{a^3 x} - a\sqrt[3]{a^2 x}}{a - \sqrt[4]{ax^3}} = \lim_{x \to a} \frac{a^{3/2}x^{1/2} - a^{5/3}x^{1/3}}{a - a^{1/4}x^{3/4}}$$

$$(LHR) = \lim_{x \to a} \frac{a^{3/2}\frac{1}{2}x^{-1/2} - a^{5/3}\frac{1}{3}x^{-2/3}}{0 - a^{1/4}\frac{3}{4}x^{-1/4}}$$

$$= \frac{\frac{1}{2}a - \frac{1}{3}a}{-\frac{3}{4}}$$

$$= -\frac{2a}{9}.$$

For the $x \to +\infty$ limit, LHR could be used again, but it is easier to notice that f satisfies $|f(x)| < Cx^{-1/4}$ for large x, so there is a horizontal asymptote $f(x) \to 0$ as $x \to \infty$. From (37.1), $f(x) < 0$ for all $x \in (0, a) \cup (a, \infty)$, so the maximum value is $f(0) = 0$.

Remark from the problem authors Finding critical points to locate maximum values can be attempted, but this is a difficult calculation and a waste of time. The formula (14.1) is a variation on an expression appearing in an early calculus book by L'Hôpital (and Johann Bernoulli) [S].

S2014-2 For integer $k \geq 1$, $f(x^{(2^k)}) = f((x^{(2^{k-1})})^2) = f(x^{(2^{k-1})})$, so by induction, $f(x^{(2^k)}) = f(x)$ for all integers $k \geq 0$. Given $\varepsilon > 0$, there is some $\delta \in (0, 1)$ and some $N \in (1, \infty)$ so that if $0 < t < \delta$ or $t > N$, then $|f(t) - f(1)| < \varepsilon$. If $0 < x < 1$, then there is some integer k such that $k > \log_2(\ln(\delta)/\ln(x))$, which is equivalent to $0 < x^{(2^k)} < \delta$, and if $x > 1$, then there is some integer k such that $k > \log_2(\ln(N)/\ln(x))$, which is equivalent to $N < x^{(2^k)}$, so in either case, $|f(x) - f(1)| = |f(x^{(2^k)}) - f(1)| < \varepsilon$. Since ε was arbitrary, $f(x) = f(1)$.

Remark from the problem authors More informal limit arguments can be attempted, but at the risk of taking some unjustified steps.

S2014-3A From $\sin = \frac{opp}{hyp}$, $\sin x = a/d$, $\sin y = b/d$, and $\sin z = c/d$, where a, b, c are the side lengths of the box and d is the long diagonal length.

$$\det(A) = \frac{1}{d^3} \det \begin{bmatrix} a & b & c \\ c & a & b \\ b & c & a \end{bmatrix}.$$

The absolute value of

$$\det \begin{bmatrix} a & b & c \\ c & a & b \\ b & c & a \end{bmatrix}$$

is the volume of a parallelepiped with side lengths all equal to d. By the scalar triple product formula, such a volume is maximized when the parallelepiped has all right angles, so it is a cube with volume d^3. The claimed inequality follows.

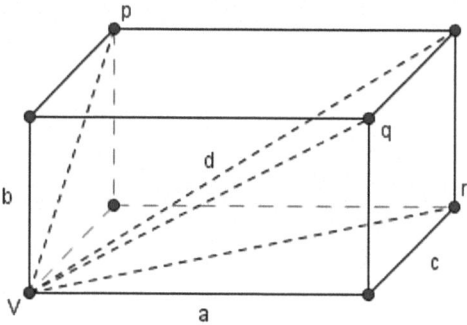

S2014-3B From $\sin = \frac{opp}{hyp}$, $\sin x = a/d$, $\sin y = b/d$, and $\sin z = c/d$, where a, b, c are the side lengths of the box and d is the long diagonal length. By the Pythagorean Theorem,

$$\sin^2(x) + \sin^2(y) + \sin^2(z) = \frac{a^2 + b^2 + c^2}{d^2} = 1$$

This gives the diagonal entries in the matrix product:

$$AA^T = \begin{bmatrix} 1 & s & s \\ s & 1 & s \\ s & s & 1 \end{bmatrix},$$

where the other entries are:

$$s = \sin(z)\sin(y) + \sin(x)\sin(z) + \sin(y)\sin(x) = \frac{bc + ac + ab}{d^2} = \frac{bc + ac + ab}{a^2 + b^2 + c^2}.$$

By the arithmetic-geometric mean inequality, $bc + ac + ab \leq \frac{1}{2}(b^2+c^2) + \frac{1}{2}(a^2+c^2) + \frac{1}{2}(a^2+b^2)$, so $0 \leq s \leq 1$. On the interval $0 \leq s \leq 1$,

$$f(s) = \det(AA^T) = \det(A)^2 = 1 - 3s^2 + 2s^3$$

is decreasing (with $f'(s) = 6s(s-1) \leq 0$) from maximum value 1 to minimum value 0, and the claimed inequality $|\det(A)| \leq 1$ follows.

S2014-3C From $\sin = \frac{opp}{hyp}$, $\sin x = a/d$, $\sin y = b/d$, and $\sin z = c/d$, where a, b, c are the side lengths of the box and d is the long diagonal length. Expanding the determinant gives

$$\det(A) = \frac{1}{d^3} \det \begin{bmatrix} a & b & c \\ c & a & b \\ b & c & a \end{bmatrix} = \frac{a^3 + b^3 + c^3 - 3abc}{d^3}.$$

The RHS is nonnegative by the three-variable arithmetic-geometric mean inequality, $\frac{1}{3}(a^3 + b^3 + c^3) \geq (a^3 b^3 c^3)^{1/3}$. Using $\sin x = a/d \leq 1$, and similarly $b/d \leq 1$, $c/d \leq 1$, the RHS also has this upper bound:

$$\frac{a^3 + b^3 + c^3 - 3abc}{d^3} = \frac{a}{d}\frac{a^2}{d^2} + \frac{b}{d}\frac{b^2}{d^2} + \frac{c}{d}\frac{c^2}{d^2} - \frac{3abc}{d^3}$$

$$\leq \frac{a^2 + b^2 + c^2}{d^2} - \frac{3abc}{d^3}$$

$$= 1 - \frac{3abc}{d^3} \leq 1.$$

So we get a stronger conclusion, $0 \leq \det(A) \leq 1 - 3\sin(x)\sin(y)\sin(z) \leq 1$.

S2014-4 Let $F(x)$ be the function $2x - 1 - \int_0^x f(t)dt$, which is continuous on $[0, 1]$ and satisfies $F(0) = -1$ and $F(1) = 1 - \int_0^1 f(t)dt \geq 1 - \int_0^1 1 dt = 0$. By the Intermediate Value Theorem, $F(x) = 0$ has at least one solution $x \in [0, 1]$. This solution is unique because F is increasing on $[0, 1]$: For $0 \leq a < b \leq 1$,

$$F(b) - F(a) = 2(b-a) - \int_a^b f(t)dt \geq 2(b-a) - 1(b-a) = b - a > 0.$$

Remark from the problem authors If $f(t)$ were continuous, then F could be proved increasing using the Fundamental Theorem of Calculus: $F'(x) = 2 - f(x) \geq 1$. However, the problem specifically omits this hypothesis.

S2014-5 Let $AB = b$, $BC = h$, $AM = y$, $MB = x$, let θ be half the angle ABC, and let α be the angle BMC. By the Law of Sines,

$$\frac{\sin\theta}{y} = \frac{\sin(\pi-\alpha)}{AC}, \frac{\sin\alpha}{h} = \frac{\sin\theta}{x} \implies \frac{\sin\alpha}{\sin\theta} = \frac{h}{x} = \frac{AC}{y} = \frac{\sqrt{b^2+h^2}}{y}.$$

We have the following system of polynomial equations:

$$x + y = b$$
$$\frac{1}{2}xh = s$$
$$bh = s + t$$
$$x^2(b^2 + h^2) = h^2 y^2.$$

Eliminating y first gives

$$x^2(b^2 + h^2) = h^2(b-x)^2 \implies x^2 b = h^2 b - 2h^2 x.$$

Multiplying both sides by h^3 gives

$$x^2 b h^3 = h^4(hb - 2hx)$$
$$(2s)^2(s+t) = h^4(s + t - 2(2s))$$
$$h = \left(\frac{4s^2(s+t)}{t - 3s}\right)^{1/4}$$
$$b = \frac{s+t}{h} = \frac{(s+t)^{3/4}(t-3s)^{1/4}}{\sqrt{2s}}.$$

The b/h ratio can be computed directly for $t = 4s$, or as

$$\frac{b}{h} = \frac{bh}{h^2} = \frac{s + 4s}{\left(\frac{4s^2(s+4s)}{4s-3s}\right)^{1/2}} = \frac{5s}{\sqrt{20s^2}} = \frac{\sqrt{5}}{2}.$$

Remark from the problem authors The equality of ratios $\frac{h}{x} = \frac{AC}{y}$ from the first step is also known as the "bisector theorem" for triangles.

S2014-6 There are eight types of students with the following populations:

$$\# \text{ RH BL boy} = 2$$
$$\# \text{ RH BR boy} = 9$$
$$\# \text{ LH BR girl} = 5$$

LH BR boy = 13 − 5 = 8
RH BR girl = 26 − 9 − 13 = 4
LH BL girl = x
RH BL girl = y
LH BL boy = z.

From equal numbers of boys and girls, $x+y+9 = 19+z$. From equal numbers of LH and RH, $x+z+13 = y+15$. From one fourth LH girls, $4(x+5) = x+y+z+28$. This is a system of three linear equations in three unknowns. Standard solution methods give the unique answer $x = 6$, $y = 7$, and $z = 3$, so the total population is $x+y+z+28 = 44$, with seven RH BL girls.

Remark from the problem authors Drawing a Venn diagram may be helpful.

S2014-7 For the first part,

$$F(e) = f(e) * f(a \cdot e) * f(a^2 \cdot e) * \ldots * f(a^{n-1} \cdot e)$$
$$= f(e) * f(a) * f(a^2) * \ldots * f(a^{n-1}),$$
$$F(a) = f(a) * f(a^2) * f(a^2 \cdot a) * \ldots * f(a^{n-2} \cdot a) * f(a^{n-1} \cdot a)$$
$$= f(a) * f(a^2) * f(a^3) * \ldots * f(a^{n-1}) * f(e) = F(e),$$

using the property that $f(e)$ commutes with other $f(g)$ at the last step. By the assumption that $a \neq e$, F is not one-to-one.

For the second part, there are lots of examples. (A correct answer must have explicit examples of G, n, a, and f.) A simple one is to let G be a two-element group $\{e, a\}$, so $n = 2$, and to define $f : G \to S_3$ by $f(e) = (12)$ and $f(a) = (23)$, or any other pair of non-commuting elements in S_3. Then $F(e) = f(e) * f(a) = (12) \circ (23) = (123)$ and $F(a) = f(a) * f(e) = (23) \circ (12) = (132)$, so F is one-to-one.

S2014-8 Let x be the number; the cube roots suggest that x may be a root of a cubic polynomial, so try expanding x^3. The following calculation has convenient cancelations and shows x satisfies $x^3 = 12 + 5x$:

$$x^3 = 6 + \sqrt{\frac{847}{27}} + 3\left((6+\sqrt{\frac{847}{27}})(36 - \frac{847}{27})\right)^{1/3}$$
$$+ 3\left((36 - \frac{847}{27})(6 - \sqrt{\frac{847}{27}})\right)^{1/3} + 6 - \sqrt{\frac{847}{27}}$$

$$= 12 + 3(36 - \frac{847}{27})^{1/3}\left((6+\sqrt{\frac{847}{27}})^{1/3} + (6-\sqrt{\frac{847}{27}})^{1/3}\right)$$

$$= 12 + 3\left(\frac{125}{27}\right)^{1/3} x = 12 + 5x.$$

The only real root of $x^3 - 5x - 12 = (x-3)(x^2 + 3x + 4)$ is $x = 3$; the conclusion is that x is rational.

Remark from the Editors This problem is similar to **P2009-4** and to ICMC **P1969-3** [AFMC].

Chapter 38
2015 Solutions

The 2015 competition was held at Taylor University. On the occasion of the 50th annual ICMC event, the Indiana Section of the MAA provided themed T-shirts to the contest participants. 2015 was also the Centennial year of the MAA.

The 2015 problem set was prepared by Professor Brian Rice at Huntington University and Professor Feng Tian at Trine University.

The **2015 problem statements** begin on page 35 in Chap. 15.

The team from Indiana—Purdue Fort Wayne, third place in 2015: L to R: Sofia Sorokina, Altun Shukurlu, Vreneli Brenneman

S2015-1

(a) The number 2015 has the super-3 representation $2015 = 3^7 - 3^5 + 3^4 - 3^2 - 3^0$.

(b) Without loss of generality we may let n be a nonnegative integer, since given a super-3 representation of n, we can find a super-3 representation of $-n$ by changing the signs of all the a_k. Now, we prove this by strong induction on n. As given in the problem statement, 0 has a super-3 representation; this is our base case. Now, let $n > 0$ and suppose that every $0 \leq x < n$ has a super-3 representation. Then n is either a multiple of 3, one more than a multiple of 3, or one less than a multiple of 3. Let q be the nearest multiple of 3 to n, and let $r = q/3$. Certainly $0 \leq r < n$, so by the inductive step it has a super-3 representation; say $r = \sum_{k=1}^{m} a_k \cdot 3^{p_k}$. Then $q = \sum_{k=1}^{m} a_k \cdot 3^{p_k+1}$; then either $n = q$, and that sum is a representation of n, or $n = q \pm 1$. In the latter case:

- If $n = q + 1$, then $n = \sum_{k=1}^{m} a_k \cdot 3^{p_k+1} + 3^0$ (so $a_{m+1} = 1$ and $p_{m+1} = 0$).

- If $n = q - 1$, then $n = \sum_{k=1}^{m} a_k \cdot 3^{p_k+1} - 3^0$ (so $a_{m+1} = -1$ and $p_{m+1} = 0$).

In any case this gives a super-3 representation of n since the exponents of 3 are all distinct. Each positive exponent is one more than its corresponding exponent in a known super-3 representation (thus all positive exponents are distinct), and if one of the exponents is 0, it is the only such exponent. This proves that n has a super-3 representation, completing the inductive step and the proof.

Remark from the problem author Several other solutions are possible, including different induction arguments and a construction based on base-3 representations.

S2015-2A Rewrite the left-hand side as

$$\sum_{k=0}^{n} \left(\binom{n}{k} 2^k \sum_{i=0}^{n-k} \binom{n-k}{i} 2^i \right).$$

Applying the binomial theorem to the interior sum, we obtain $(2+1)^{n-k} = 3^{n-k}$, so we find that the left-hand side is equal to

$$\sum_{k=0}^{n} \binom{n}{k} 2^k 3^{n-k}.$$

Applying the binomial theorem again, we find that the left-hand side is equal to $(2+3)^n = 5^n$ as required.

S2015-2B The term inside the double sum counts the number of ways to choose, from a group of n people, two disjoint committees A and B of sizes k and i, respectively, and then to choose a (possibly empty) third committee C from among the members of both committees A and B. The double sum then adds all these up over all possible values of k and i. Hence the sum as a whole counts the number of ways to choose from n people two disjoint committees A and B, and then a third committee C from among the members of A and B. We can count this number in a different way: Each person can be either (1) on no committees at all, (2) on committee A only, (3) on committee B only, (4) on committees A and C, or (5) on committees B and C. Since each person independently has five possible assignments, the total number of ways to do this is 5^n, as required.

S2015-3 Since $\overline{AB}$ and $\overline{CD}$ intersect, the quadrilateral $ACBD$ is planar and inscribed in the circle that its plane cuts from the sphere. By the power of a point theorem, $AX \cdot BX = CX \cdot DX$, whence $1 \cdot 4 = 2 \cdot DX$ and $DX = 2$. Since $\overline{AB}$ and $\overline{CD}$ are perpendicular, the area N of $ACBD$ is $N = \frac{1}{2}AB \cdot CD = \frac{1}{2}(5 \cdot 4) = 10$.

The same argument shows that $1 \cdot 4 = AX \cdot BX = EX \cdot FX = 3 \cdot FX$, and thus $FX = \frac{4}{3}$. Since $\overline{EF}$ is perpendicular to the plane containing $ACBD$, the octahedron is divided by that plane into two right pyramids with base area $N = 10$ and heights $EX = 3$ and $FX = \frac{4}{3}$, respectively. These two pyramids have volumes $\frac{1}{3}EX \cdot N = \frac{1}{3}(3 \cdot 10) = 10$ and $\frac{1}{3}FX \cdot N = \frac{1}{3}(\frac{4}{3} \cdot 10) = \frac{40}{9}$. Hence the volume of the octahedron is $10 + \frac{40}{9} = \frac{130}{9}$.

Remark from the problem author The argument above can be used to prove the little-known fact that the power of a point theorem is true in three dimensions as well as in two.

Remark from the Editors At the 2015 ICMC event, at least one team's submitted solution included a verification of the pyramid volume formula by calculating an integral.

S2015-4

(a) The minimum number of trials is 3. In the first trial, turn on switches 1, 2, 3, and 4; in the second, switches 1, 2, 5, and 6, and in the third, 1, 3, 5, and 7. Then the light that was on in all three trials is that controlled by switch 1, that which is on in trials 1 and 2, but not trial 3, is controlled by switch 2, and so on. It is easy to verify that each light is on in a difference subset of the trials, so this determines which switches control which lights.

(b) Observe that if two switches x and y are always on or off together in our trials, then our trials cannot distinguish which of them control which light. For instance, if in every trial, switches 1 and 2 are either both on or both off, and these switches control lights A and B, we will not be able to determine whether switch 1 controls light A or light B. Therefore, if we can use a set of trials to determine which switches control which lights, it must be the case that the function from switches to subsets of tests given by $f(x) = \{\alpha \mid \text{Light } x \text{ is on in Test } \alpha\}$ is one-to-one. Since the number of subsets of all trials is 2 to the power of the number of trials, and the number of switches is 8, it follows that there must be at least three trials.

Remark from the problem author The same argument can be used to show that in the general case of n switches, we need to use at least $\lceil \log_2(n) \rceil$ trials to determine which switches control which lights.

S2015-5

(a) Let $a_1, a_2, \ldots$ be an infinite sequence of zeros of f in the interval $[0, 1]$. Since every bounded infinite sequence has a convergent subsequence, we may find such a convergent subsequence $b_1, b_2, \ldots$. Let $x = \lim_{n \to \infty} b_n$. Certainly, $x \in [0, 1]$. Now, there are either infinitely many b_n less than x or infinitely many b_n greater than x; without loss of generality, suppose the former. It then follows that we may choose an infinite increasing sequence $c_1, c_2, \ldots$ from among the b_n; simply begin with $c_1 < x$, and continue choosing c_{n+1} from among the b_i, such that $c_n < c_{n+1} < x$. (We can always do this because otherwise we would have infinitely many b_i which are less than c_n, which is impossible since $\lim_{n \to \infty} b_n = x > c_n$.) Since this increasing sequence is a subsequence of the b_n, it necessarily converges to x.

Since f is differentiable, it is continuous, and it follows that $f(x) = f(\lim_{n \to \infty} c_n) = \lim_{n \to \infty} f(c_n) = 0$. Now by Rolle's Theorem, there is a zero of f' between c_n and c_{n+1} for every $n \geq 1$; let $c_n^{(1)}$ be a zero of f' between c_n and c_{n+1}. Then the sequence $c_1^{(1)}, c_2^{(1)}, \ldots$ is an increasing sequence of zeros of f', and by the squeeze theorem it has limit x. As f' is itself differentiable and thus continuous, it follows as above that $f'(x) = 0$. Continuing by induction, we find that for every n, there is an increasing sequence $c_1^{(n)}, c_2^{(n)}, \ldots$ of zeros of $f^{(n)}$ with limit x, and hence that $f^{(n)}(x) = 0$.

This shows that our chosen x satisfies the required condition.

(b) One such function is given by

$$f(x) = \begin{cases} \sin\left(\frac{1}{x}\right) \exp\left(-\frac{1}{x^2}\right) & x \neq 0 \\ 0 & x = 0. \end{cases}$$

Remark from the problem author The function given above for the second part is a modification of a classic example showing that a nonzero function can have identically zero Taylor series at 0.

Remark from the Editors The differentiability of the function $\exp(-1/x^2)$ is considered in ICMC **P1982-4** [AFMC].

S2015-6 Let G be a k-color-planar graph. Each of the k sets of edges with the same color forms a planar graph on the vertices of G. By the Four Color Theorem, each of these graphs can be 4-colored. Doing so, each vertex of G gets a k-tuple of colors corresponding to its color in each of the k planar graphs. If we assign a different color to each k-tuple, this gives us a 4^k-coloring of G (since adjacent vertices must differ in at least one coordinate as they are adjacent in some one of the k planar graphs). Thus, every k-color-planar graph can be 4^k-colored (so $c_k = 4^k$ works).

Remark from the problem author Is this value $c_k = 4^k$ best possible? We do not know and would be interested in a proof either way.

S2015-7

(a) Because $*$ is associative, we may omit parentheses without any ambiguity. Let $a, b \in S$ be arbitrary. We have

$$a * b = a * (a * a * b * a)$$
$$= a * a * a * (a * a * b * a) * a$$
$$= a * (a * a * a * a) * b * a * a$$
$$= a * a * b * a * a$$
$$= b * a,$$

where the assumption that $a*a*b*a = b$ is used in several steps (including one step where $a*a*a*a = a$, the $b = a$ case). This shows that $*$ is commutative.

(b) It follows that, for any $a, b \in S$, $b * (a * a * a) = (a * a * a) * b = b$, so $(a * a * a)$ acts as an identity for $*$. The identity is unique, since if e and f are both identities, then $e = e * f = f$. Call the identity e. Furthermore, since $a * a * a = e$ for every $a \in S$, it follows that $a * a$ is an inverse for a. Thus, $*$ is an associative binary operation with an identity and inverses, and therefore $(S, *)$ is a group.

Moreover, the fact that $a * a * a = e$ for every $a \in S$ implies that the order of every element of S divides 3. It follows from Cauchy's Theorem that $|S|$ cannot be divisible by any prime other than 3, so $|S|$ is a power of 3. (Recall Cauchy's Theorem states that if G is a finite group and $|G|$ is a multiple of a prime number p, then there exists an element of G with order p.)

Remark from the problem author In the last step, the fundamental theorem of finite abelian groups can be used in place of Cauchy's theorem.

S2015-8 Define $g(x) = f(x) - x^{\frac{3}{2}}$. Then $g(0) = 0$ and $g(1) = 0$, so by the Mean Value Theorem, there is some $c \in (0, 1)$ such that $g'(c) = 0$ and hence $f'(c) = \frac{3}{2}\sqrt{c}$. Thus $(f'(c))^2 = \frac{9}{4}c$. Applying the Mean Value Theorem to $(f')^2$ on the interval $[0, c]$, we obtain that there is some $a \in (0, c)$ (and hence in $(0, 1)$) such that $2f'(a)f''(a) = ((f')^2)'(a) = \frac{9}{4}$. Thus $f'(a)f''(a) = \frac{9}{8}$ as required.

Chapter 39
2016 Solutions

The 51st competition was held at Franklin College. 2016 was also a Centennial year—the preliminary organization of the Indiana Section of the MAA occurred at the October 26–28, 1916, meeting of the Mathematics Section of the Indiana State Teachers' Association. However, with a delay possibly related to global events including the war and a pandemic, the first meeting of the MAA Indiana Section did not occur until October 16, 1924 [Da].

The 2016 problem set was prepared by Professors Paul Fonstad, Justin Gash, and Stacy Hoehn at the host institution.

The **2016 problem statements** begin on page 39 in Chap. 16.

S2016-1 To prove that the set is infinite, we will use induction to show that for $a \in A$, $n(a)|a$ whenever a consists of 3^m ones (i.e., $n(a) = 3^m$) for any integer $m \geq 0$. For the base case when $m = 0$, note that if $n(a) = 3^0 = 1, a = 1$, so $n(a)|a$. Now assume that for some integer $k \geq 0$, $n(b)$ divides b when b is the element of A consisting of 3^k ones, and consider the element c of A consisting of 3^{k+1} ones. Note that $b \cdot 10^{3^k}$ consists of 3^k ones followed by 3^k zeros, while $b \cdot 10^{2 \cdot 3^k}$ consists of 3^k ones followed by $2 \cdot 3^k$ zeros. Therefore, c can be built from three copies of b via the formula

$$c = b + b \cdot 10^{3^k} + b \cdot 10^{2 \cdot 3^k}$$
$$= b(1 + 10^{3^k} + 10^{2 \cdot 3^k}).$$

Note that $3^k = n(b)$ divides b by the inductive hypothesis, and 3 divides $d = (1 + 10^{3^k} + 10^{2 \cdot 3^k})$ since the sum of the digits of d is 3. Therefore, $3^k \cdot 3 = 3^{k+1} = n(c)$ divides c, which completes the proof.

S2016-2A f must be the identity function. In other words, $f(n) = n$ for all $n \in \mathbb{Z}$.

Step 1. We will use strong induction to show that $f(n) = n$ for all integers $n \geq 2$.

For the base case, note that $f(2) = 2$ is property (i) of f. Now assume that for some integer $k \geq 2$, $f(n) = n$ for all integers $2 \leq n \leq k$, and consider $f(k+1)$.

- Case 1: If $k+1$ is composite, then $k+1$ can be written as the product of two integers m and n where $2 \leq m \leq k$ and $2 \leq n \leq k$. Then

$$f(k+1) = f(mn)$$
$$= f(m)f(n) \text{ by property (ii) of } f$$
$$= mn \text{ by the inductive hypothesis}$$
$$= k+1.$$

- Case 2: If $k+1$ is prime, then $k+2$ must be composite since $k \geq 2$, so $k+2$ can be written as the product of two integers m and n, where $2 \leq m \leq k+1$ and $2 \leq n \leq k+1$. We can actually lower the upper bound on both m and n from $k+1$ to k since $\frac{k+2}{k+1} < 2$ when $k \geq 2$. Then similarly to above, $f(k+2) = f(mn) = f(m)f(n) = mn = k+2$. Since $f(k) = k$ (by the inductive hypothesis) and $f(k+2) = k+2$, we then know that $f(k+1) = k+1$ by property (iii) of f.

So far we have shown that $f(n) = n$ for all integers $n \geq 2$.
Step 2. Note that

$$f(0) = f(2 \cdot 0) = f(2)f(0) = 2f(0),$$

so it must be the case that $f(0) = 0$. Since $f(0) = 0$ and $f(2) = 2$, we know that $f(1) = 1$ by property (iii) of f.
Step 3. $f(-1) = -1$ because $f(-1) < f(0) = 0$ and

$$(f(-1))^2 = f(-1)f(-1) = f((-1) \cdot (-1)) = f(1) = 1.$$

To complete the proof, we need to show that $f(n) = n$ for negative integers $n \leq -2$. Note that if n is a negative integer, then $-n$ is a positive integer, so we know that $f(-n) = -n$ by our previous work. Then

$$f(n) = f((-1) \cdot (-n)) = f(-1) \cdot f(-n) = (-1) \cdot (-n) = n.$$

Thus, for all integers n, $f(n) = n$, so f must be the identity function.

S2016-2B The abovementioned Step 2 could instead be established by using property (ii) to note that $f(0) \cdot f(0) = f(0 \cdot 0) = f(0)$ and $f(1) \cdot f(1) = f(1 \cdot 1) = f(1)$, so $f(0) = 0$ or 1 and $f(1) = 0$ or 1. Then by property (iii), $f(0) = 0 < f(1) = 1$.

Then, the abovementioned Step 1 can be given a different proof by induction. Suppose $f(n) = n$ for $1 \leq n \leq 2^{2^k}$. The $k = 0$ case is the previously established

$f(1) = 1$ and property (i), $f(2) = 2$. For $k > 0$, $f(2^{2^{(k+1)}}) = f(2^{2^k} \cdot 2^{2^k}) = f(2^{2^k}) \cdot f(2^{2^k}) = 2^{2^k} \cdot 2^{2^k} = 2^{2^{(k+1)}}$. Because f is strictly increasing, the only possible f on the interval from $2^{2^k} + 1$ to $2^{2^{(k+1)}} - 1$ is the identity function.

S2016-3 One such solution would be the points

$(0, 0, 0)$	$(1, 0, 0)$
$(\cos(2\pi/5), \sin(2\pi/5), 0)$	$(\cos(4\pi/5), \sin(4\pi/5), 0)$
$(\cos(6\pi/5), \sin(6\pi/5), 0)$	$(\cos(8\pi/5), \sin(8\pi/5), 0)$
$(0, 0, 1)$	$(0, 0, -1)$.

Geometrically, these points represent the origin, the five vertices of a regular pentagon in the xy-plane with each point distance 1 away from the origin, and the two points on the line perpendicular to the xy-plane through the origin distance 1 away from the origin. To see that any choice of three points gives you an isosceles triangle, consider the following cases:

Case 1: One of the three points is the origin. Then by construction the other two points are each distance one away from the origin, and so two sides of the triangle are congruent. Note that when the other two points are $(0, 0, 1)$ and $(0, 0, -1)$, the triangle is degenerate.

Case 2: All three points are on the pentagon. Then either one point is a vertex adjacent to the other two vertices on the pentagon, or one point is a vertex opposite to the other two vertices on the pentagon. In either case, the distance between that vertex and the other vertices is the same, giving us again an isosceles triangle.

Case 3: One of the vertices is outside the xy-plane, the two other vertices are on the pentagon. Then by the Pythagorean theorem, both points on the pentagon are distance $\sqrt{2}$ away from the third vertex, which gives us an isosceles triangle.

Case 4: One vertex is on the pentagon, and the other two are off the xy-plane. Then the point on the plane is $\sqrt{2}$ away from the points off the plane, which again gives us an isosceles triangle.

S2016-4

(a) Define $h : A \times [0, 1] \to A$ by $h(a, t) = (1 - t)f(a) + tg(a) = (1 - t)a^2 + ta^3$. Note that $h(a, t) \in A = [0, 1]$ for all $a \in A = [0, 1]$ and all $t \in [0, 1]$. Moreover, h is a continuous function such that $h(a, 0) = (1-0)f(a) + 0g(a) = f(a)$ and $h(a, 1) = (1 - 1)f(a) + 1g(a) = g(a)$, so f and g are homotopic.

(b) • *Reflexivity*: Suppose $f : A \to A$ is any continuous function on A. Define $h : A \times [0, 1] \to A$ by $h(a, t) = f(a)$. Then h is a continuous function such that $h(a, 0) = f(a)$ and $h(a, 1) = f(a)$ for all $a \in A$, so f is homotopic to f.

• *Symmetry*: Suppose $f, g : A \to A$ are any two continuous functions on A such that f is homotopic to g. This implies that there exists a continuous function $h : A \times [0, 1] \to A$ such that $h(a, 0) = f(a)$ and $h(a, 1) = g(a)$ for all $a \in A$. Define $H : A \times [0, 1] \to A$ by $H(a, t) = h(a, 1-t)$. Then H is a continuous function such that $H(a, 0) = h(a, 1 - 0) = h(a, 1) = g(a)$ and $H(a, 1) = h(a, 1 - 1) = h(a, 0) = f(a)$ for all $a \in A$. Therefore, g is homotopic to f.

- *Transitivity*: Suppose that $f, g, i : A \to A$ are any three continuous functions on A such that f is homotopic to g and g is homotopic to i. This implies that there exist continuous functions $h : A \times [0, 1] \to A$ and $h' : A \times [0, 1] \to A$ such that $h(a, 0) = f(a), h(a, 1) = g(a), h'(a, 0) = g(a)$, and $h'(a, 1) = i(a)$ for all $a \in A$. Since $h(a, 1) = g(a) = h'(a, 0)$ for all $a \in A$, the function $H : A \times [0, 1] \to A$ defined by

$$H(a, t) = \begin{cases} h(a, 2t) & \text{if } 0 \leq t \leq \frac{1}{2} \\ h'(a, 2t - 1) & \text{if } \frac{1}{2} \leq t \leq 1 \end{cases}$$

is continuous, and $H(a, 0) = h(a, 2 \cdot 0) = h(a, 0) = f(a)$ and $H(a, 1) = h'(a, 2 \cdot 1 - 1) = h'(a, 1) = i(a)$ for all $a \in A$. Therefore, f is homotopic to i.

S2016-5

(a) $T : \mathbb{R}^2 \to \mathbb{R}^2$ given by $T(x, y) = (y, x + y)$ is a linear transformation such that $T(f_n, f_{n+1}) = (f_{n+1}, f_n + f_{n+1}) = (f_{n+1}, f_{n+2})$ for all integers $n \geq 0$. We will prove using induction that $T^n(0, 1) = (f_n, f_{n+1})$ for all integers $n \geq 1$. For the base case, note that $T^1(0, 1) = T(0, 1) = (1, 0 + 1) = (1, 1) = (f_1, f_2)$. Now assume that $T^k(0, 1) = (f_k, f_{k+1})$ for some integer $k \geq 1$, and consider $T^{k+1}(0, 1)$. Note that

$$T^{k+1}(0, 1) = T(T^k(0, 1))$$
$$= T(f_k, f_{k+1}) \text{ by the inductive hypothesis}$$
$$= (f_{k+1}, f_k + f_{k+1})$$
$$= (f_{k+1}, f_{k+2}).$$

Thus, we can conclude that $T^n(0, 1) = (f_n, f_{n+1})$ for all integers $n \geq 1$.

(b) Since $T(1, 0) = (0, 1 + 0) = (0, 1)$ and since $T(0, 1) = (1, 0 + 1) = (1, 1)$, the standard matrix A of T is

$$A = \begin{bmatrix} 0 & 1 \\ 1 & 1 \end{bmatrix}.$$

The characteristic polynomial of A is $(0 - \lambda)(1 - \lambda) - 1 = \lambda^2 - \lambda - 1$, so the eigenvalues of A are $\lambda_1 = \frac{1+\sqrt{5}}{2}, \lambda_2 = \frac{1-\sqrt{5}}{2}$. Note that $\lambda_1 = \varphi$ is the golden ratio, while $\lambda_2 = 1 - \varphi$.

(c) The vector $\mathbf{v}_1 = \begin{bmatrix} 1 \\ \varphi \end{bmatrix}$ is an eigenvector corresponding to $\lambda_1 = \varphi = \frac{1+\sqrt{5}}{2}$ since

$$A\mathbf{v}_1 = \begin{bmatrix} \varphi \\ 1+\varphi \end{bmatrix} = \varphi \begin{bmatrix} 1 \\ 1+\frac{1}{\varphi} \end{bmatrix} = \varphi \begin{bmatrix} 1 \\ \varphi \end{bmatrix} = \varphi \mathbf{v}_1,$$

and similarly, $\mathbf{v}_2 = \begin{bmatrix} 1 \\ 1-\varphi \end{bmatrix}$ is an eigenvector corresponding to $\lambda_2 = 1 - \varphi$. Note that

$$\begin{bmatrix} 0 \\ 1 \end{bmatrix} = \frac{1}{\sqrt{5}} (\mathbf{v}_1 - \mathbf{v}_2).$$

Therefore,

$$A^n \begin{bmatrix} 0 \\ 1 \end{bmatrix} = \frac{1}{\sqrt{5}} \left(\varphi^n \mathbf{v}_1 - (1-\varphi)^n \mathbf{v}_2 \right)$$

$$= \frac{1}{\sqrt{5}} \left(\varphi^n \begin{bmatrix} 1 \\ \varphi \end{bmatrix} - (1-\varphi)^n \begin{bmatrix} 1 \\ 1-\varphi \end{bmatrix} \right)$$

$$= \frac{1}{\sqrt{5}} \begin{bmatrix} \varphi^n - (1-\varphi)^n \\ \varphi^{n+1} - (1-\varphi)^{n+1} \end{bmatrix}.$$

Since A is the standard matrix of T, this implies that

$$T^n(0, 1) = \frac{1}{\sqrt{5}} \left(\varphi^n - (1-\varphi)^n, \varphi^{n+1} - (1-\varphi)^{n+1} \right).$$

By part (a), the first coordinate of $T^n(0, 1)$, namely $\frac{1}{\sqrt{5}} (\varphi^n - (1-\varphi)^n)$, equals f_n. This provides us our desired non-recursive expression for f_n:

$$f_n = \frac{1}{\sqrt{5}} \left(\varphi^n - (1-\varphi)^n \right). \tag{39.1}$$

Remark from the Editors Of course (39.1) is not the only possible non-recursive expression, and elementary properties of φ such as $1 - \varphi = -1/\varphi$ lead to equivalent formulas. If one considers matrix exponentiation (rather than real number exponentiation) to be a non-recursive expression, then another solution using the above matrix A is the product $f_n = \begin{bmatrix} 1 & 0 \end{bmatrix} \begin{bmatrix} 0 & 1 \\ 1 & 1 \end{bmatrix}^n \begin{bmatrix} 0 \\ 1 \end{bmatrix}$ (or a similar expression in terms of the matrix Q from **P2003-6** in Chap. 3).

S2016-6

(a) Let a, b be any two real numbers with $a < b$. By the Mean Value Theorem, there exists a real number c such that $a < c < b$ and

$$f'(c) = \frac{f(b) - f(a)}{b - a}.$$

Since $f'(c) = 0$, this implies that $0 = \frac{f(b) - f(a)}{b - a}$, so that $f(a)$ must equal $f(b)$. Since a and b are arbitrary real numbers, f must be a constant function.

(b) To see that this is false, suppose that $A = (0, 1) \cup (2, 3)$, and define $f : A \to \mathbb{R}$ by

$$f(a) = \begin{cases} 1 & \text{if } 0 < a < 1, \\ 2 & \text{if } 2 < a < 3. \end{cases}$$

Then $f'(a) = 0$ for all a in A, but f is not a constant function.

S2016-7

(a) Assume to the contrary that a and b are two elements of such a group G such that $a \neq b$ but $a^3 = b^3$. Note that $(b^3)^{-1} = (b^{-1})^3$ because

$$b^3(b^{-1})^3 = (bb^{-1})^3 = e^3 = e.$$

Therefore, $a^3 = b^3$ implies that

$$a^3(b^3)^{-1} = b^3(b^3)^{-1}$$

$$a^3(b^{-1})^3 = e$$

$$(ab^{-1})^3 = e \text{ by property (i) of } G.$$

Since there are no elements of G of order 3 (by property (ii) of G) and $(ab^{-1})^3 = e$, it must be the case that $ab^{-1} = e$, which implies that $a = b$, which contradicts our assumption that $a \neq b$. Therefore, it must be the case that $a^3 \neq b^3$ whenever $a \neq b$.

(b) Assume that G contains n elements and that $\phi : G \to G$ is defined by $\phi(g) = g^3$. Note that the previous part implies that ϕ is one-to-one since $\phi(a) = a^3$ equals $\phi(b) = b^3$ if and only if $a = b$. We will show that ϕ must be onto via contradiction. Assume that ϕ is not onto. Then there exists $g \in G$ such that g is not in the image of ϕ. Then the image of the n elements of G must lie inside the set $G \setminus \{g\}$ that only contains $(n - 1)$ elements. Then the pigeonhole principle implies that there must exist some $g' \in G \setminus \{g\}$ and two distinct elements a, b in G such that $\phi(a) = g' = \phi(b)$. However, this is impossible since ϕ is one-to-one. Therefore, it must be the case that ϕ is onto. Since ϕ is one-to-one and onto, ϕ is a bijection.

S2016-8 Consider, for example, the following sample preference table for 27 voters in an election with 5 candidates: A, B, C, D, and E:

Rank/number of votes	2	6	7	8	4
First	C	D	B	A	C
Second	E	C	E	D	B
Third	D	E	C	E	E
Fourth	A	B	D	C	D
Fifth	B	A	A	B	A

- Plurality—A wins with the most first place votes (eight of them)
- Plurality with Elimination—no candidate has more than half the first place votes, so we eliminate the candidate(s) with the fewest first place votes. Candidate E is eliminated. The new preference table is

Rank/number of votes	2	6	7	8	4
First	C	D	B	A	C
Second	D	C	C	D	B
Third	A	B	D	C	D
Fourth	B	A	A	B	A

Still, no candidate has more than half the first place votes, so we eliminate the candidate(s) with the fewest. Candidates C and D are eliminated. The new preference table is

Rank/number of votes	2	6	7	8	4
First	A	B	B	A	B
Second	B	A	A	B	A

B now has more than half of the first place votes, so B is the winner.
- Borda count—the point totals are as follows:
    - A: 2(2)+6(1)+7(1)+8(5)+4(1)=61
    - B: 2(1)+6(2)+7(5)+8(1)+4(4)=73
    - C: 2(5)+6(4)+7(3)+8(2)+4(5)=91
    - D: 2(3)+6(5)+7(2)+8(4)+4(2)=90
    - E: 2(4)+6(3)+7(4)+8(3)+4(3)=90

    Therefore, C wins.

- Pairwise Comparison—the relevant matchups are as follows:
  - D vs. A: D wins 19–7 because 19 voters like D better than A.
  - D vs. B: D wins 16–11 because 16 voters like D better than B.
  - D vs. C: D wins 14–13 because 14 voters like D better than C.
  - D vs. E: D wins 14–13 because 14 voters like D better than E.

  Therefore, since D went undefeated and untied and since all other candidates lost at least once (to D), D wins.
- Survivor—we eliminate the candidate(s) with the most last place votes. Candidate A is eliminated. The new preference table is

Rank/number of votes	2	6	7	8	4
First	C	D	B	D	C
Second	E	C	E	E	B
Third	D	E	C	C	E
Fourth	B	B	D	B	D

Now B is eliminated, so the new preference table is

Rank/number of votes	2	6	7	8	4
First	C	D	E	D	C
Second	E	C	C	E	E
Third	D	E	D	C	D

Now D is eliminated, so the new preference table is

Rank/number of votes	2	6	7	8	4
First	C	C	E	E	C
Second	E	E	C	C	E

Now C is eliminated, so E wins.

Remark from the Editors Considering the length and complexity of this problem, one strategy during a timed competition would be to aim for partial credit by constructing a table comparing only three or four of the voting methods instead of all five.

Chapter 40
2017 Solutions

At the 52nd ICMC event at Earlham College, the organizers loaned each team a straightedge and compass.

The 2017 problem set was prepared by Professor Joshua Cole at St. Joseph's College and Professor Stacy Hoehn at Franklin College.

The **2017 problem statements** begin on page 45 in Chap. 17.

S2017-1

(a) Draw a circle centered at A that passes through B, as well as a circle centered at B that passes through A. These two circles will intersect in two points which we will call C and D. Then draw the line segment connecting C and D. This line segment will intersect $\overline{AB}$ in a point that we will call M. M is the midpoint of $\overline{AB}$. See the figure below.

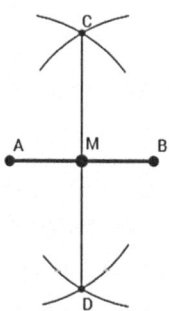

(b) Use the straightedge to extend $\overline{AB}$ to the left. Then draw the circle centered at A that passes through B. This circle will intersect $\overleftrightarrow{AB}$ at B and another point. Draw a circle centered at this new point that passes through B, as well as a circle centered at B that passes through this new point. Then draw the line segment that connects the two points where these circles intersect. This line segment passes through A and is perpendicular to $\overline{AB}$. The corner D of our square will be one of the points where this line segment intersects the circle centered at A that passes through B. Finally, draw a circle centered at D that passes through A and a circle centered at B that passes through A. These two circles will intersect at A and another point, namely the final corner C of our square. To complete the square, draw the line segments connecting C with D and C with B. See the figure.

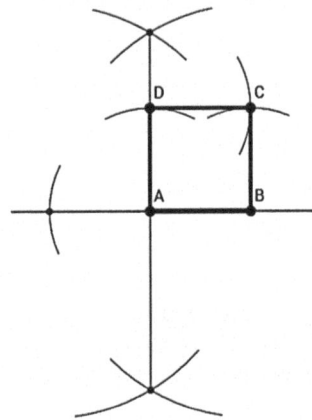

(c) Let s denote the length of $\overline{AB}$. Since triangle ABC is a right isosceles triangle, the length of $\overline{AC}$ is then $\sqrt{2}s$.

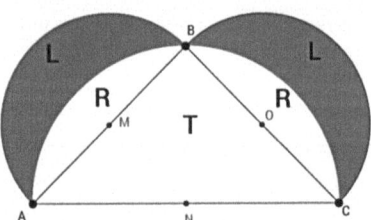

If T, R, and L denote the areas of the indicated regions in the above figure, $T = \frac{1}{2}s^2$ since it is the area of a triangle with base s and height s,

$$T + 2R = \frac{1}{2}\pi \left(\frac{\sqrt{2}s}{2}\right)^2 = \frac{\pi}{4}s^2$$

since $T + 2R$ is the area inside a semicircle with radius $\frac{\sqrt{2}s}{2}$, and

$$L + R = \frac{1}{2}\pi \left(\frac{s}{2}\right)^2 = \frac{\pi}{8}s^2$$

since $L+R$ is the area inside a semicircle with radius $\frac{s}{2}$. We can use the first two equations to see that $R = \frac{\pi}{8}s^2 - \frac{1}{4}s^2$. We can then use this value for R as well as the third equation given above to see that L, which is the area of the lune, is $L = \frac{1}{4}s^2$. Thus, the area of the lune will equal the area of a square that has side length $\frac{1}{2}s$, which is just the length of the line segment from B to the midpoint of $\overline{AB}$. Therefore, to construct a square with area equal to the area of one of the lunes, we can use our construction from part (a) to construct the midpoint M of $\overline{AB}$ and then use our construction from part (b) to construct a square that has $\overline{MB}$ as one of its sides. (Alternatively, since triangle ABC is a right isosceles triangle, the polygon that connects B, the midpoint M of $\overline{AB}$, the midpoint N of $\overline{AC}$, and the midpoint O of $\overline{BC}$ will be a square with side length $\frac{1}{2}s$.)

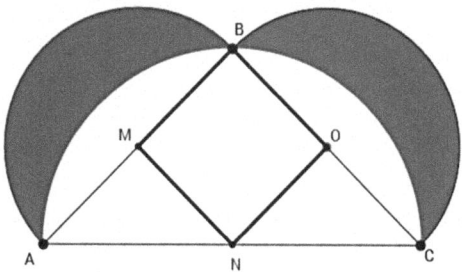

S2017-2 This series does converge. To see this, we make a comparison to the series $\sum_{n=1}^{\infty} \left(\frac{1}{2}\right)^n$, which is a convergent geometric series, since its ratio is $\frac{1}{2} < 1$. We demonstrate that $0 \le (a_n)^n \le \left(\frac{1}{2}\right)^n$ for all n. If n is odd, then a_n is the nth partial sum of the geometric series $\sum_{n=1}^{\infty} \left(\frac{1}{3}\right)^n$, which converges to $\frac{\frac{1}{3}}{1-\frac{1}{3}} = \frac{\frac{1}{3}}{\frac{2}{3}} = \frac{1}{2}$. Since the terms of that series are nonnegative, it follows that $0 \le a_n \le \frac{1}{2}$. Thus $0 \le (a_n)^n \le \left(\frac{1}{2}\right)^n$, as we desire for the comparison test. If n is even, then $u_n = |\sin(n)\cos(n)| = \frac{1}{2}|\sin(2n)|$. Since $|\sin(2n)| \le 1$ for all n, $0 \le a_n \le \frac{1}{2}$. Then $0 \le (a_n)^n \le \left(\frac{1}{2}\right)^n$ for all n, as we desire for the comparison test. We have shown that for all n (odd or even), $0 \le (a_n)^n \le \left(\frac{1}{2}\right)^n$. Hence, by the comparison test, $\sum_{n=1}^{\infty}(a_n)^n$ converges.

S2017-3 Note that $x^2 + ax + b = 0$ has at least one real root as long as the discriminant, which is $a^2 - 4b$ in this case, is nonnegative. Thus, we need to find the probability that $a^2 - 4b \ge 0$, i.e., $b \le \frac{1}{4}a^2$. Since each student is picking a number from $[0, 1]$ according to the uniform probability distribution, finding the desired probability is equivalent to finding the proportion of $[0, 1] \times [0, 1]$, where $b \le \frac{1}{4}a^2$.

This region is shaded in the figure below, where the horizontal axis corresponds to a and the vertical axis corresponds to b.

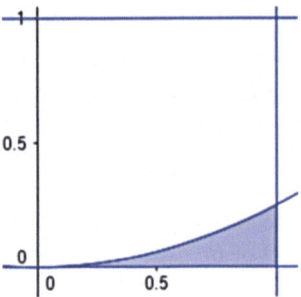

The area of the shaded region is

$$\int_0^1 \frac{1}{4}a^2 \, da = \frac{1}{12},$$

while the area of the entire square is 1. Thus, the desired probability is the ratio of these areas, namely $\frac{1}{12}$.

S2017-4

(a) Suppose that the diagram below represents a 3×3 multiplicative magic square with magic product P.

a	b	c
d	e	f
g	h	i

Then $P = abc = def = ghi = adg = beh = cfi = aei = ceg$.

$$P^4 = (aei)(ceg)(def)(beh) = (abc)(def)(ghi)e^3 = P^3 e^3,$$

and the conclusion is $P = e^3$.

Remark from the Editors Part (a) repeats **P2003-4**; see **S2003-4** in Chap. 26 for an alternate solution.

(b) By the previous part, we know that the magic product must be a perfect cube. Moreover, the magic product must have at least nine distinct divisors because each entry in the magic square is a divisor of the magic product. The first five perfect cubes, 1, 8, 27, 64, and 125, have 1, 4, 4, 7, and 4 divisors, respectively. The next perfect cube, 216, has 16 divisors, so it is a candidate for the minimal magic product. Since the sample array given below is a 3 × 3 multiplicative magic square with magic product 216, 216 is, in fact, the minimal magic product.

12	1	18
9	6	4
2	36	3

Other multiplicative magic squares with magic product 216 are possible. However, our solution to the previous part implies that all of them will have a 6 in the middle square.

S2017-5

(a) This formula relies on the fact that for any positive integer k, the sum of the integers from 1 to k is $\frac{1}{2}k(k+1)$. We will first consider the case when n is odd. In this situation,

$$s(n) = 0 + 1 + 1 + 2 + 2 + \cdots + \frac{n-1}{2} + \frac{n-1}{2}$$

$$= 2\left(1 + 2 + \cdots + \frac{n-1}{2}\right)$$

$$= 2\left(\frac{1}{2}\left(\frac{n-1}{2}\right)\left(\frac{n-1}{2}+1\right)\right)$$

$$= \left(\frac{n-1}{2}\right)\left(\frac{n+1}{2}\right)$$

$$= \frac{n^2 - 1}{4}.$$

When n is even,

$$s(n) = 0 + 1 + 1 + 2 + 2 + \cdots + \left(\frac{n}{2} - 1\right) + \left(\frac{n}{2} - 1\right) + \frac{n}{2}$$

$$= \left(1 + 2 + \cdots + \left(\frac{n}{2} - 1\right)\right) + \left(1 + 2 + \cdots + \left(\frac{n}{2} - 1\right) + \frac{n}{2}\right)$$

$$= \left(\frac{1}{2}\left(\frac{n}{2} - 1\right)\left(\frac{n}{2}\right)\right) + \left(\frac{1}{2}\left(\frac{n}{2}\right)\left(\frac{n}{2} + 1\right)\right)$$

$$= \left(\frac{1}{2}\left(\frac{n-2}{2}\right)\left(\frac{n}{2}\right)\right) + \left(\frac{1}{2}\left(\frac{n}{2}\right)\left(\frac{n+2}{2}\right)\right)$$

$$= \left(\frac{n^2 - 2n}{8}\right) + \left(\frac{n^2 + 2n}{8}\right)$$

$$= \frac{n^2}{4}.$$

Thus,

$$s(n) = \begin{cases} \frac{n^2-1}{4} & \text{if } n \text{ is odd,} \\ \frac{n^2}{4} & \text{if } n \text{ is even.} \end{cases}$$

(b) Note that $m+n$ and $m-n$ differ by $2n$, which is even. Thus, $m+n$ and $m-n$ are either both odd or both even. In the case when they are both odd,

$$s(m+n) - s(m-n) = \frac{(m+n)^2 - 1}{4} - \frac{(m-n)^2 - 1}{4}$$
$$= \frac{(m^2 + 2mn + n^2 - 1) - (m^2 - 2mn + n^2 - 1)}{4}$$
$$= \frac{4mn}{4}$$
$$= mn.$$

Meanwhile, in the case when they are both even,

$$s(m+n) - s(m-n) = \frac{(m+n)^2}{4} - \frac{(m-n)^2}{4}$$
$$= \frac{(m^2 + 2mn + n^2) - (m^2 - 2mn + n^2)}{4}$$
$$= \frac{4mn}{4}$$
$$= mn.$$

S2017-6 We will give a proof by induction. For the base case, we will consider the case when $n = 1$. Then $A = (1)$ or $A = (-1)$, so $\det(A)$ equals 1 or -1, both of which are divisible by $2^{n-1} = 2^0 = 1$. For the inductive hypothesis, assume that for some $k \geq 1$, the determinant of any $k \times k$ matrix whose entries are all ± 1 is divisible by 2^{k-1}. For the inductive step, suppose that A is any $(k+1) \times (k+1)$ matrix such that every entry of A is ± 1. Let B be the $(k+1) \times (k+1)$ matrix obtained from A by replacing row 1 of A by the sum of row 1 with row 2. Since each entry of A is 1 or -1, the entries in the first row of B must all be 2, 0, or -2, while the entries in the remaining rows of B must all be 1 or -1. Moreover, $\det(A) = \det(B)$ since the type of elementary row operation performed does not change the determinant. Therefore, it suffices to show that $\det(B)$ is divisible by $2^{(k+1)-1} = 2^k$. By using cofactor expansion along the first row of B, we see that

$$\det(B) = \sum_{j=1}^{k+1}(-1)^{1+j}b_{1j}\det(M_{1j}),$$

where M_{1j} is the $k \times k$ matrix obtained from B by removing row 1 and column j. For each j, M_{1j} is a $k \times k$ matrix whose entries are all 1 or -1, so by the inductive hypothesis, $\det(M_{1j})$ is divisible by 2^{k-1} for all j. Moreover, for each j, b_{1j} is divisible by 2 since b_{1j} equals 2, 0, or -2. Thus, for each j, $b_{1j}\det(M_{1j})$ is divisible by $2 \cdot 2^{k-1} = 2^k$, so

$$\det(B) = \sum_{j=1}^{k+1}(-1)^{1+j}b_{1j}\det(M_{1j})$$

is also divisible by 2^k. Since $\det(A) = \det(B)$, $\det(A)$ is divisible by 2^k, completing our induction.

S2017-7 Let σ denote the permutation implemented by the robot. Then in cycle notation,

$$\sigma^7 = (A\ 7\ J\ 9\ 10\ 5\ Q\ 6\ 3\ K\ 8\ 4\ 2).$$

Thus, $|\sigma^7| = 13$. We claim that this implies that $|\sigma| = 13$ as well. To see this, note that $|\sigma^7| = 13$ implies that $(\sigma^7)^{13} = e$, where e denotes the identity permutation. Then $\sigma^{91} = e$. Thus it implies that $|\sigma|$ divides 91, meaning that $|\sigma|$ is 1, 7, 13, or 91. But $|\sigma|$ cannot equal 1 or 7 because then σ^7 would have to equal e, which it does not. Moreover, if $|\sigma|$ was 91, then the least common multiple of the lengths of the cycles when σ is written in disjoint cycle notation would have to be 91, which would mean that σ would have disjoint cycles of lengths 7 and 13 or a cycle of length 91, both of which are impossible with only a total of 13 cards to permute. Therefore, σ must have order 13. Since $\sigma^{13} = e$, $(\sigma^7)^2 = \sigma^{14} = \sigma\sigma^{13} = \sigma e = \sigma$. Hence, in cycle notation,

$$\sigma = (\sigma^7)^2 = (A\ J\ 10\ Q\ 3\ 8\ 2\ 7\ 9\ 5\ 6\ K\ 4),$$

so the order of the cards after the first shuffle is

$$4, 8, Q, K, 9, 5, 2, 3, 7, J, A, 10, 6.$$

S2017-8 First note that A cannot be finite. For a finite set A, around each element a of A, there is an open interval small enough so that it contains a but no other element of A, contradicting the given condition on A. Thus, A must be infinite.

Next, we will show that A cannot be countably infinite via contradiction. Suppose that A is countably infinite. Then the elements of A can be listed out as $a_1, a_2, a_3, \ldots$. Let U_1 be the open interval $(a_1 - 1, a_1 + 1)$ around a_1. By the condition on A, this open interval must contain another element in A; without loss of generality, assume that a_2 is in U_1 (if not, relabel the elements of A). Next, let U_2 be a small enough open interval around a_2 so that the closure of U_2 (i.e., the closed interval consisting of U_2 and the corresponding endpoints) is contained inside U_1 and so that x_1 is not in the closure of U_2. Again, by the condition on A, U_2 must contain another element of A; without loss of generality, assume that a_3 is in U_2. Now choose an open interval U_3 around a_3 such that the closure of U_3 is contained in U_2 and so that neither a_1 nor a_2 is contained in the closure of U_3. Without loss of generality, assume that a_4 is in U_3. In general, for each $n \geq 2$, find a small enough open interval U_n around a_n such that:

- The closure of U_n is contained in U_{n-1}.
- $a_1, a_2, \ldots, a_{n-1}$ are not contained in the closure of U_n.
- a_{n+1} is contained in U_n (after a possible relabeling of the points in $\{a_{n+1}, a_{n+2}, \ldots\}$).

For each $n \geq 1$, let $V_n = \text{closure}(U_n) \cap A$. Note that each V_n is closed and bounded, hence compact. Moreover, $V_1 \supset V_2 \supset V_3 \supset \cdots$. Let $V = \bigcap_{i=1}^{\infty} V_n$. Since V is the intersection of a decreasing nested sequence of non-empty compact subsets of A, V is a non-empty subset of A. However, none of the a_n's can be in V since for $i \geq n$, a_n is not in V_i. Since we have arrived at a contradiction, A cannot be countably infinite. Thus, it must be the case that A is uncountably infinite.

Chapter 41
2018 Solutions

The 53rd contest, temporarily renamed the Intersectional Collegiate Mathematics Competition, was held at Valparaiso University, during the Tri-Section meeting of the Illinois, Indiana, and Michigan Sections of the Mathematical Association of America. Over 50 teams, from all three states, participated.

The 2018 problem set was prepared by Professors Daniel Maxin, Zsuzsanna Szaniszlo, and Tiffany Kolba at the host institution.

The **2018 problem statements** begin on page 49 in Chap. 18.

S2018-1 For the claimed equality to happen, one needs $|\sin(jx)| = 1$ for every $j = 1, \ldots, n$. Since $n \geq 2$, we have

$$|\sin(x)| = 1 \text{ and } |\sin(2x)| = 1,$$

which leads to $x = \frac{\pi}{2} + k\pi$ and $2x = \frac{\pi}{2} + \ell\pi$ with k, ℓ integers. Combining the two, one obtains $\ell - 2k = \frac{1}{2}$, which is a contradiction.

S2018-2 Rationalizing the denominators, one obtains the following telescopic summation:

$$(\sqrt{2} - 1) + (\sqrt{3} - \sqrt{2}) + \cdots + (\sqrt{n+1} - \sqrt{n}) \geq 100,$$

which simplifies to

$$\sqrt{n+1} - 1 \geq 100.$$

Hence the smallest n is 10200.

S2018-3 Consider the function

$$f(x) = a_1^x + a_2^x + \cdots + a_n^x.$$

Notice that $f(0) = n$ and, from the hypothesis, $f(x) \geq f(0) = n$. Therefore $f(x)$ has a local minimum at $x = 0$. From Fermat's theorem, $f'(0) = 0$. But,

$$f'(x) = a_1^x \ln(a_1) + \cdots + a_n^x \ln(a_n).$$

Then

$$f'(0) = \ln(a_1) + \cdots + \ln(a_n) = \ln(a_1 a_2 \cdots a_n) = 0$$

and $a_1 a_2 \cdots a_n = 1$.

S2018-4 First, we use the following row and column operations, which will not change the value of the determinant: Add rows 2 and 3 to row 1. Then, subtract column 1 from column 2 and column 1 from column 3. With these operations, the determinant becomes

$$\det \begin{bmatrix} x+y+z & 0 & 0 \\ y & x-y & z-y \\ z & y-z & x-z \end{bmatrix}$$

$$= (x+y+z) \det \begin{bmatrix} 1 & 0 & 0 \\ y & x-y & z-y \\ z & y-z & x-z \end{bmatrix}$$

$$= (x+y+z)(x^2 + y^2 + z^2 - xy - yz - zx)$$

$$= \frac{1}{2}(x+y+z)((x-y)^2 + (y-z)^2 + (z-x)^2) \geq 0.$$

S2018-5 Notice

$$\frac{1}{x_{k+1}} = \frac{1}{x_k} - \frac{1}{x_k + 1}, \quad \text{or} \quad \frac{1}{x_k + 1} = \frac{1}{x_k} - \frac{1}{x_{k+1}}.$$

Adding up for $k = 1, \ldots, 100$, we get a telescoping cancelation and

$$S_{100} = \frac{1}{x_1} - \frac{1}{x_{101}} = 2 - \frac{1}{x_{101}}.$$

Since x_n is increasing and $x_3 > 1$, we have $0 < \frac{1}{x_{101}} < 1$ and $S_{100} = 2 - \frac{1}{x_{101}} < 2$. So $\lfloor S_{100} \rfloor = 1$.

S2018-6A First notice that, by dividing by $|AB|^2$, the equality to prove becomes

$$\frac{|AM|}{|AB|}\frac{|AD|}{|AB|} + \frac{|BM|}{|AB|}\frac{|BE|}{|AB|} = 1.$$

Consider the perpendicular from M to AB and denote by P the intersection of this perpendicular with AB.

Notice that the triangles AMP and ABD are similar and the ratio of the segments opposite the congruent angles is the same. Hence

$$\frac{|AM|}{|AB|} = \frac{|AP|}{|AD|}.$$

Analogously, notice that the triangles BMP and BAE are also similar and, from there, we infer that

$$\frac{|BM|}{|AB|} = \frac{|PB|}{|BE|}.$$

Substituting these into the equality to prove, we obtain

$$\frac{|AP|}{|AD|}\frac{|AD|}{|AB|} + \frac{|PB|}{|BE|}\frac{|BE|}{|AB|} = \frac{|AP|+|PB|}{|AB|} = \frac{|AB|}{|AB|} = 1.$$

S2018-6B Using the fact that the inscribed triangles $\triangle ABD$ and $\triangle ABE$ have right angles at D and E,

$$|AM||AD| + |BM||BE|$$
$$= |AM|(|AM| + |BM|\cos\angle BMD) + |BM|(|BM| + |AM|\cos\angle AME)$$
$$= |AM|^2 + |BM|^2 + 2|AM||BM|\cos\angle AME$$
$$= |AM|^2 + |BM|^2 - 2|AM||BM|\cos\angle AMB$$
$$= |AB|^2.$$

The last step uses the Law of Cosines.

S2018-7 Let $Y =$ the number of coin flips. On average, it takes 2 flips to observe a head. Hence, on average, it takes $2X$ flips to observe X heads. Then, we have

$$E[Y] = E[E[Y|X]] = E[2X] = 2E[X] = 2(3.5) = 7.$$

For a more direct computation, note that all outcomes of the die roll have probability $\frac{1}{6}$. For each of those equally weighted cases, one would expect to flip a coin 2, 4, 6, 8, 10, and 12 times, respectively. Thus,

$$E[Y]$$
$$= \frac{1}{6} \times 2 + \frac{1}{6} \times 4 + \frac{1}{6} \times 6 + \frac{1}{6} \times 8 + \frac{1}{6} \times 10 + \frac{1}{6} \times 12$$
$$= \frac{1}{6}(2 + 4 + 6 + 8 + 10 + 12)$$
$$= 7.$$

S2018-8 Let $c \geq 0$ be the number of clubs. Since each pair of clubs (but not three) shares a member, the number of pairs of clubs cannot be more than the number of people. In other words, $\binom{c}{2} \leq n$. This condition becomes

$$c^2 - c - 2n \leq 0,$$

which is a quadratic in c, concave upward with two real roots, one positive and one nonpositive. Therefore c must be bounded by the positive root, i.e.,

$$0 \leq c \leq (1 + \sqrt{1 + 8n})/2. \tag{41.1}$$

Claim: The number of clubs that may be formed can be any integer satisfying (41.1). In particular, the maximum number, $[(1 + \sqrt{1 + 8n})/2]$ (where $[x]$ denotes the integer part of $x > 0$), can be achieved.

Note that for any number ($n \geq 0$) of townspeople, a town with no clubs ($c = 0$) or one club ($c = 1$) meets, vacuously, the criteria of the problem, and it is also permitted that there are exactly two clubs ($c = 2$) as long as there is at least one person who is a member of both ($n \geq 1$).

To establish the claim for any $c \geq 3$ satisfying (41.1), we assume there are c clubs and need to show that there exists an assignment of members to clubs satisfying the criteria. Because $c(c-1)/2 \leq n$, there exists a one-to-one function f from the list of $c(c-1)/2$ pairs of clubs to the list of $n \geq 3$ people. For a pair of clubs, $\{c_j, c_k\}$, $j \neq k$, if $f(\{c_j, c_k\}) = A$, then assign person A to be a member of c_j and of c_k, and no other clubs. If person B is not in the image of f, then B belongs to no clubs. Now each club c_k has at least one member $f(\{c_j, c_k\})$ in common with any other club c_j, and by construction there is no person who is a member of more than two clubs.

For example, if $n = 6$, then the maximum is $c = 4$. Let the people be denoted by A, B, C, D, E, F, the clubs by 1, 2, 3, 4. Based on the assignment:

A in 1, 2 B in 1, 3 C in 1, 4 D in 2, 3 E in 2, 4 F in 3, 4,

the members of the clubs are as follows:

$$1: A, B, C \quad 2: A, C, E \quad 3: B, C, F \quad 4: D, E, F.$$

For another example, if $n = 5$, then the maximum is $c = [(1+\sqrt{41})/2] = 3$. Let the people be denoted by A, B, C, D, E, the clubs by 1, 2, 3. Based on the assignment:

$$A \text{ in } 1, 2 \quad B \text{ in } 1, 3 \quad C \text{ in } 2, 3,$$

the members of the clubs are as follows:

$$1: A, B \quad 2: A, C \quad 3: B, C,$$

and D and E do not belong to any club. This meets the criteria. If there were a fourth club, then the number of pairs of clubs would be 6 as in the previous example, so by the first criterion, one of the five people must belong to two distinct pairs of clubs—either three or four clubs, a violation of the second criterion.

Chapter 42
2019 Solutions

The 54th competition was held at the University of Indianapolis.

The 2019 problem set was prepared by Professor Paul Fonstad at Franklin College.

The **2019 problem statements** begin on page 51 in Chap. 19.

S2019-1A

(a) Let $f(x) = x + \sin(x)$ and $g(x) = -x$. Then f and g are both continuous. Individually, each function is monotone, but the sum $(f+g)(x) = \sin(x)$ is not.

(b) The function

$$f(x) = \frac{\sin(\frac{1}{x})}{x}$$

is not bounded in any neighborhood of $x = 0$, and neither of the one-sided limits at $x = 0$ exists.

(c) Consider the function

$$f(x) = \begin{cases} 0 & \text{if } x \text{ is rational,} \\ x & \text{if } x \text{ is irrational.} \end{cases}$$

This $f(x)$ is continuous at $x = 0$ but discontinuous everywhere else.

(d) The function $f(x) = x^3$ has $f'(0) = 0$, yet the function is increasing on $\mathbb{R}$.

(e) If $f(x) = 1/x$ on the domain $(-\infty, 0) \cup (0, \infty)$, then $F(x) = \ln|x|$ is an antiderivative of f, satisfying $F'(x) = f(x)$ at every point in the domain. F is defined at both $x = -1$ and $x = 1$, but $\int_{-1}^{1} \frac{1}{x} dx$ does not converge.

S2019-1B For part (e), there is another counterexample with $f(x)$ defined on all of $\mathbb{R}$ and $F(x)$ satisfying $F'(x) = f(x)$ at every point, but where $\int_{-1}^{1} f(x)dx$ fails to exist as a proper Riemann integral because f is unbounded near $x = 0$.

$$F(x) = \begin{cases} |x|^{3/2} \sin(1/x) & \text{for } x \neq 0, \\ 0 & \text{for } x = 0. \end{cases}$$

$$f(x) = \begin{cases} \frac{3}{2}|x|^{1/2} \sin(1/x) - |x|^{-1/2} \cos(1/x) & \text{for } x > 0, \\ 0 & \text{for } x = 0, \\ -\frac{3}{2}|x|^{1/2} \sin(1/x) - |x|^{-1/2} \cos(1/x) & \text{for } x < 0. \end{cases}$$

The integral $\int_{-1}^{1} f(x)dx$ could be evaluated using some other method (an improper integral, or some method of integration other than Riemann's) to give $F(1) - F(-1)$.

S2019-2 All such triangles must pass through the point $(0, 4)$. To see this, first note that since the right angle occurs at the origin, the legs of the right triangle are formed by perpendicular lines whose equations can be written as $y = mx$ and $y = -x/m$ for some $m > 0$. These lines intersect the parabola at the points $(4m, 4m^2)$ and $(-4/m, 4/m^2)$, respectively. Thus, the equation for the hypotenuse of the triangle is

$$y = \left(\frac{4m^2 - \frac{4}{m^2}}{4m + \frac{4}{m}}\right)(x - 4m) + 4m^2$$

$$= \left(\frac{4m^2 - \frac{4}{m^2}}{4m + \frac{4}{m}}\right)x + 4.$$

No matter what m is, when x equals 0, y will equal 4. Thus, the hypotenuse of the right triangle will always pass through the point $(0, 4)$.

S2019-3

(a) Note that the amount of money that party guest n makes (or loses) is given by $f(n) = 10(e_n - o_n)$, where e_n denotes the number of even positive divisors of n and o_n denotes the number of odd positive divisors of n (including 1). Thus, if n is odd, guest n will lose money since $e_n = 0$. If n is even but not divisible by 4, guest n will break even, since every even divisor of n can be paired with an odd divisor of n. Finally, if n is divisible by 4, party guest n will make money since n will have more even divisors than odd divisors since every odd divisor is paired with an even divisor but not every even divisor is paired with an odd divisor. Thus, 30 guests will lose money, 15 will break even, and 15 will make money.

(b) By the previous part, we only need to consider the 15 guests whose number n is divisible by 4. The party guest who makes the most money will correspond to the number n for which the number of even divisors exceeds the number of odd divisors by the most. If n is a power of 2, $n = 2^k$, then $f(n) = 10(k-1)$, for a maximum of $f(32) = 40$. When n has a prime factorization of the form $2^2 p$, for p an odd prime (3, 5, 7, 11, 13), $f(n) = 10(4-2) = 20$. Similarly, when n has a prime factorization of the form $2^3 p$ (for $p = 3, 5, 7$), $f(n) = 10(6-2) = 40$. The only numbers n remaining are $f(36) = f(2^2 3^2) = 10(6-3) = 30$, $f(48) = f(2^4 3^1) = 10(8-2) = 60$, and $f(60) = f(2^2 3^1 5^1) = 10(8-4) = 40$. Thus, the 48th party guest in line will make the most money.

(c) By part (a), we only need to consider guests whose number n is odd. The party guest who loses the most money will correspond to the number n which has the most odd divisors. $f(1) = 10(0-1) = -10$. Considering odd primes $1 < p < q$, $f(p) = 10(0-2) = -20$, $f(p^2) = 10(0-3) = -30$, and $f(pq) = 10(0-4) = -40$. The only numbers n remaining are $f(27) = f(3^3) = 10(0-4) = -40$ and $f(45) = f(3^2 5^1) = 10(0-6) = -60$. So, the guest who lost the most would be the 45th guest in line.

(d) The sum of positive f values from part (b) is

$$(10 + 20 + 30 + 40) + 20 \cdot 5 + 40 \cdot 3 + 30 + 60 + 40 = 450.$$

The sum of negative f values from part (c) is

$$-10 - 20 \cdot 16 - 30 \cdot 3 - 40 \cdot 8 - 40 - 60 = -840.$$

Since $\sum_{n=1}^{60} f(n) = -390$, Two-Face and his henchmen made $390. Note that by part (a), you only need to consider those values of n that are odd or that are divisible by 4 when computing this sum.

Remark from the Editors This problem and solution are similar to ICMC **P1996-3** [AFMC].

S2019-4

(a) Let X denote the number of flips needed to get your first tail. Flipping a H on your first toss adds one to your required number of tosses and gets you no closer to your goal. Meanwhile, flipping a T on your first toss completes your task in a single toss. Thus,

$$E(X) = P(H)E(X|H) + P(T)E(X|T)$$
$$= P(H)(E(X) + 1) + P(T)(1)$$

$$= \frac{1}{2}(E(X)+1) + \frac{1}{2}.$$

This implies $\frac{1}{2}E(X) = 1$, which implies $E(X) = 2$.
Alternatively,

$$E(X) = \sum_{n=1}^{\infty} n \left(\frac{1}{2}\right)^n = 2.$$

(b) Let Y denote the number of flips needed to achieve your first string of heads followed by tails. Flipping a T on your first toss adds one to your required number of tosses and gets you no closer to your goal. If you flip a H on your first toss, you are just waiting on your first T. Let X denote the number of additional tosses required to get your first T if your first toss is H. From the previous part, we know that $E(X) = 2$. Then

$$E(Y) = P(T)E(Y|T) + P(H)(E(X)+1)$$
$$= \frac{1}{2}(E(Y)+1) + \frac{1}{2}(2+1)$$
$$= \frac{1}{2}(E(Y)) + 2.$$

This implies $\frac{1}{2}E(Y) = 2$, which implies $E(Y) = 4$.
Alternatively,

$$E(Y) = \sum_{n=1}^{\infty} n \left(\frac{n-1}{2^n}\right) = 4.$$

(c) Let Z denote the number of flips needed to achieve the first string of heads followed by heads. Flipping a T on your first toss adds one to your number of required flips but does not get you closer to observing two heads. Likewise, starting out by flipping HT adds two flips to your total but does not get you closer to observing two heads. Meanwhile, starting out with HH takes two flips and completes the task in two tosses. Thus,

$$E(Z) = P(T)E(Z|T) + P(HT)E(Z|HT) + P(HH)E(Z|HH)$$
$$= P(T)(E(Z)+1) + P(HT)(E(Z)+2) + P(HH)(2)$$
$$= \frac{1}{2}(E(Z)+1) + \frac{1}{4}(E(Z)+2) + \frac{1}{4}(2).$$

This implies $\frac{1}{4}E(Z) = \frac{3}{2}$, which implies $E(Z) = 6$.

S2019-5 Claim: For any integer n, $A^n = \begin{bmatrix} 1 & 2019n \\ 0 & 1 \end{bmatrix}$.

We will prove this using induction, twice. For the base cases, first observe that $A^1 = \begin{bmatrix} 1 & 2019 \cdot 1 \\ 0 & 1 \end{bmatrix}$, $A^0 = I = \begin{bmatrix} 1 & 2019 \cdot 0 \\ 0 & 1 \end{bmatrix}$, and that $A^{-1} = \begin{bmatrix} 1 & 2019 \cdot (-1) \\ 0 & 1 \end{bmatrix}$ since

$$\begin{bmatrix} 1 & 2019 \\ 0 & 1 \end{bmatrix} \begin{bmatrix} 1 & -2019 \\ 0 & 1 \end{bmatrix} = \begin{bmatrix} 1 & 0 \\ 0 & 1 \end{bmatrix}.$$

Next, assume that for some positive integer k, $A^k = \begin{bmatrix} 1 & 2019k \\ 0 & 1 \end{bmatrix}$. Then

$$A^{k+1} = A^k A = \begin{bmatrix} 1 & 2019k \\ 0 & 1 \end{bmatrix} \cdot \begin{bmatrix} 1 & 2019 \\ 0 & 1 \end{bmatrix}$$

$$= \begin{bmatrix} 1 \cdot 1 + 2019k \cdot 0 & 1 \cdot 2019 + 2019k \cdot 1 \\ 0 \cdot 1 + 1 \cdot 0 & 0 \cdot 2019 + 1 \cdot 1 \end{bmatrix}$$

$$= \begin{bmatrix} 1 & 2019(k+1) \\ 0 & 1 \end{bmatrix}.$$

Finally, assume that for some negative integer k, $A^k = \begin{bmatrix} 1 & 2019k \\ 0 & 1 \end{bmatrix}$. Then, similarly,

$$A^{k-1} = A^k A^{-1} = \begin{bmatrix} 1 & 2019k \\ 0 & 1 \end{bmatrix} \begin{bmatrix} 1 & -2019 \\ 0 & 1 \end{bmatrix}$$

$$= \begin{bmatrix} 1 & 2019(k-1) \\ 0 & 1 \end{bmatrix}.$$

This completes our proof.

S2019-6 The sequences $(s_n)_{n=1}^{\infty}$ and $(t_n)_{n=1}^{\infty}$ must converge, while not enough information is provided to determine whether the sequence $(u_n)_{n=1}^{\infty}$ converges.

To see that $(s_n)_{n=1}^{\infty}$ must converge, note that it is a monotone decreasing sequence that is bounded below by t_1, since for all $n \in \mathbb{N}$, $s_n \geq t_n \geq t_1$ by hypothesis. Every monotone decreasing sequence that is bounded below converges.

To see that $(t_n)_{n=1}^{\infty}$ must converge, note that it is a monotone increasing sequence that is bounded above by s_1, since for all $n \in \mathbb{N}$, $t_n \leq s_n \leq s_1$ by hypothesis. Every monotone increasing sequence that is bounded above converges.

Not enough information is given to determine whether the sequence $(u_n)_{n=1}^{\infty}$ will converge. There are some examples that satisfy all the given criteria for which

$(u_n)_{n=1}^\infty$ will converge, and others for which it will not converge. For example, if $s_n = 1 + \frac{1}{n}$, $t_n = -1 - \frac{1}{n}$, and $u_n = 0$, then $(s_n)_{n=1}^\infty$ is a monotone decreasing sequence, $(t_n)_{n=1}^\infty$ is a monotone increasing sequence, and $s_n \geq u_n \geq t_n$ for all $n \in \mathbb{N}$, and $(u_n)_{n=1}^\infty$ does converge. Meanwhile, for the same $(s_n)_{n=1}^\infty$ and $(t_n)_{n=1}^\infty$, the sequence defined by $u_n = (-1)^n$ also satisfies the given conditions, but $(u_n)_{n=1}^\infty$ does not converge.

S2019-7 Since x and y generate G, every element of G can be written as a finite product of x's, y's, x^{-1}'s, and y^{-1}'s.

Step 1. Based on the relation $xx = e$, we have $x^{-1} = x$, and based on the relation $yyyy = e$, we have $y^{-1} = y^3$, so every element of G can be written as a finite product of just x's and y's, where you never need more than one x in a row or more than three y's in a row.

Step 2. We next show that any expression (already simplified as in Step 1.) with three y's in a row followed or preceded by an x can be rewritten using the $xyxyxy = e$ relation to have a smaller number of y's, without increasing its total number of letters:

$$y^3 x = y^{-1} x^{-1} = (xy)^{-1} = xyxy.$$

Similarly, using $xyxyxy = e \implies yxyxy = x \implies yxyxyx = e$,

$$xy^3 = x^{-1} y^{-1} = (yx)^{-1} = yxyx.$$

Then, any expression with three blocks of y's separated by two x's (of the form $\cdots yxy^k xy \cdots$) can also be simplified to have a smaller number of y's and a smaller number of letters:

$$yxyxy = x,$$
$$yxy^2 xy = (yxyx)(xyxy) = (xy^3)(y^3 x) = xy^2 x,$$
$$yxy^3 xy = y(yxyx)xy = y^2 xy^2.$$

Any product in G has a minimum length expression in x, y as in Step 1, and among products of the same length, we can consider words with the minimum number of y's. By the above calculations, such a minimal word does not have three consecutive y's unless it is y^3, and does not have three blocks of y's separated by two x's. This leaves the following list (possibly with repeats) of possible minimal expressions: There are four expressions of the form $y^{\{0,1,2,3\}}$, three of the form $xy^{\{0,1,2\}}$, twelve of the form $x^{\{0,1\}} y^{\{1,2\}} xy^{\{0,1,2\}}$, and eight of the form $x^{\{0,1\}} y^{\{1,2\}} xy^{\{1,2\}} x$. The conclusion so far is that G has at most 27 elements.

Step 3. If $y = e$, then $G = \{e, x\}$ and G has at most two elements. If $y^2 = e$, then the list from Step 2. becomes: $G = \{e, y, xy^{\{0,1\}}, x^{\{0,1\}}yxy^{\{0,1\}}, x^{\{0,1\}}yxyx\}$, with at most 10 elements. If $y^3 = e = y^4$, then $y = e$. The remaining case is that G has a subgroup with exactly four elements, $\{0, y, y^2, y^3\}$, and the number of elements of G is a multiple of 4. In any case, by the upper bound from Step 2., G has at most 24 elements.

Step 4. An example showing that 24 is the maximum that can be achieved is the symmetric group on four elements, generated by $x = (12)$ and $y = (1234)$.

One way to visualize groups is via a Cayley graph that illustrates the relations between the generators x and y. A Cayley graph for this 24-element group is the following truncated octahedron:

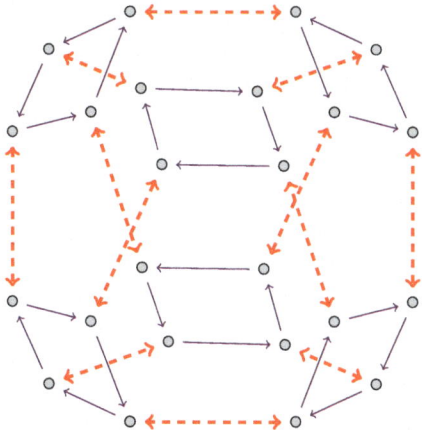

S2019-8 The faces of the pyramid T are determined by the xy-plane, the xz-plane, the yz-plane, and the plane P defined by $z = -2x - 3y + 24$. Since one corner of the rectangular prism R must be at the origin, the volume of R will be maximized when its faces meeting at the origin are also aligned with the coordinate planes, and the opposite corner is on the plane P. If this corner is located at the point $(x, y, -2x - 3y + 24)$ on the plane P, then the volume of R is given by

$$V(x, y) = xy(-2x - 3y + 24) = -2x^2 2y - 3xy^2 + 24xy.$$

Taking partial derivatives, we get that

$$V_x(x, y) = -4xy - 3y^2 + 24y = y(-4x - 3y + 24)$$

and that

$$V_y(x, y) = -2x^2 - 6xy + 24x = x(-2x - 6y + 24).$$

Setting these partial derivatives equal to zero and solving for x and y, we see that $(x, y) = (4, \frac{8}{3})$ is one critical point of V. Using the second partial derivative test, we can see that this critical point corresponds to a maximum for V. This means that the maximum volume would be

$$V = 4(\frac{8}{3})(-2(4) - 3(\frac{8}{3}) + 24) = 4(\frac{8}{3})(8) = 85\frac{1}{3}$$

cubic units.

Chapter 43
2021 Solutions

By the spring of 2021 during the pandemic, faculty and students had begun to adapt to working in virtual and remote environments. The Spring 2021 Indiana MAA Section meeting was conducted virtually as a one-day Zoom meeting, and the 55th ICMC happened on an earlier day, organized into a decentralized, remote format, where each team would meet on its own campus, with a local faculty proctor. After the teams completed the exam in the allotted amount of time, the proctor scanned and emailed their solutions to the Section's Student Activities Coordinator who distributed them to graders across the state. The results were announced during the Zoom meeting.

The 2021 problem set was prepared by Professor Harold Reiter at the University of North Carolina at Charlotte.

The **2021 problem statements** begin on page 57 in Chap. 21.

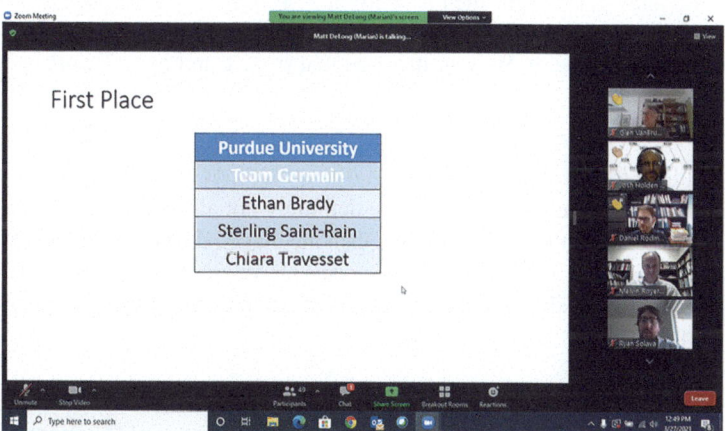

Applause via Zoom for the 2021 competition's first place team from Purdue University West Lafayette

S2021-1

(a) Notice that since A is upper triangular, $A^n = \begin{bmatrix} 1 & 2021*n \\ 0 & 1 \end{bmatrix}$. Hence $A^{2021} = \begin{bmatrix} 1 & 2021^2 \\ 0 & 1 \end{bmatrix}$.

(b) We can reverse the previous problem to see that B can also be chosen to be upper triangular and that $B = \begin{bmatrix} 1 & 1 \\ 0 & 1 \end{bmatrix}$ works.

S2021-2 The polynomial $q(x)$ must factor into the form $(cx+d)^2(ex+f)^2$ because it has a double root at each point where $mx + b$ is tangent to the graph of $p(x)$. Suppose that r and s are these double roots. We use results about the roots, due to Viète. The sum of the roots, $2r + 2s$, must equal $-(32/16) = 2$. Thus $s = 1 - r$. The sums of the products of the roots, $r*r + r*s + s*r + s*s = (1-r)^2 + r^2 + 4r(1-r)$, must equal $-104/16$. Solving this for r yields $r = -3/2$, so $s = 5/2$. We can then find m and b. The product of the roots equals $(232 - b)/16$, which yields $b = 7$. For m, we can simply evaluate $p'(r)$ or $p'(s)$, yielding $m = 2$.

S2021-3

(a) Let $P(n)$ be the product of the digits. If any digit is even, then $P(n)$ is even. Since n is also a multiple of 5, it must be a multiple of 10. But then $P(n) = 0$, a contradiction.

(b) The condition above that n must be a multiple of 5, but not a multiple of 10, means that the one's place digit must be 5. If a is the hundred's place digit and b is the ten's place digit, then $n = 100a + 10b + 5 = 5*5ab$. Dividing by 5, $20a + 2b + 1 = 5ab$. Rearranging, $5ab - 20a - 2b - 1 = 0$. Factoring, $(5a - 2)(b - 4) = 9$. We can then see that $a = 1$ and $b = 7$ work, yielding $n = 175$.

S2021-4 Let $AB = A - B$, etc. Then $||AB||^2 = 72$, $||AC||^2 = 24$, and $||BC||^2 = 48$. This shows that the given vertices are definitely not all on the same face of the supposed cube. If x is the edge of a cube, then the diagonal of a face is $x\sqrt{2}$ and the diagonal of the cube is $x\sqrt{3}$. The squared norms above are in a $3:2:1$ ratio, which shows that AC is the edge of the cube, BC is the diagonal of a face, and AB is the diagonal of the cube.

S2021-5

(a) Start with 1, and pair it with any of the seven remaining numbers. Continue with the smallest number not already used, and pair it with any of the five numbers remaining. Continuing, we see that the number of these partitions is $7*5*3*1 = 105$.

(b) Define

$$N(P) = \sum_{i=1}^{4} |x_i - y_i|^2, \qquad (43.1)$$

so that

$$N(P) + 2V(P) = \left(\sum_{i=1}^{4}(x_i - y_i)^2\right) + 2\left(\sum_{i=1}^{4} x_i y_i\right)$$

$$= \sum_{i=1}^{4}(x_i^2 + y_i^2) = \sum_{k=1}^{8} k^2$$

is constant (not depending on the choice of partition P). So, $V(P)$ is maximized whenever $N(P)$ is minimized; this occurs when the differences $|x_i - y_i|$ are as small as possible. The partition

$$P_1 = \{\{1, 2\}, \{3, 4\}, \{5, 6\}, \{7, 8\}\}$$

minimizes these differences and yields the value $V(P_1) = 100$. On the other hand, maximizing $N(P)$ is achieved by making the differences as large as possible. The partition

$$P_2 = \{\{1, 8\}, \{2, 7\}, \{3, 6\}, \{4, 5\}\}$$

does this, yielding a value $V(P_2) = 60$.

To check that $N(P_2) = 84$ is in fact the maximum, and that the above "greedy" partition P_2 is the unique partition achieving it, consider the following Lemma: Given any four real numbers in strictly increasing order: $x_1 < x_2 < x_3 < x_4$, define N_4 on partitions in analogy with (43.1) by

$$N_4\left(\{\{x_i, x_j\}, \{x_k, x_\ell\}\}\right) = |x_j - x_i|^2 + |x_\ell - x_k|^2;$$

then $\{\{x_1, x_4\}, \{x_2, x_3\}\}$ is the unique partition maximizing N_4.

Proof of Lemma Compare with the only other two partitions into pairs:

$$N_4(\{\{x_1, x_4\}, \{x_2, x_3\}\}) - N_4(\{\{x_1, x_2\}, \{x_3, x_4\}\})$$
$$= (x_4 - x_1)^2 + (x_3 - x_2)^2 - (x_1 - x_2)^2 - (x_3 - x_4)^2$$
$$= 2x_1 x_2 + 2x_3 x_4 - 2x_1 x_4 - 2x_2 x_3 \qquad (43.2)$$

$$= 2(x_4 - x_2)(x_3 - x_1) > 0,$$

$$N_4\left(\{\{x_1, x_4\}, \{x_2, x_3\}\}\right) - N_4\left(\{\{x_1, x_3\}, \{x_2, x_4\}\}\right)$$
$$= (x_4 - x_1)^2 + (x_3 - x_2)^2 - (x_3 - x_1)^2 - (x_4 - x_2)^2$$
$$= 2x_1 x_3 + 2x_2 x_4 - 2x_1 x_4 - 2x_2 x_3 \tag{43.3}$$
$$= 2(x_4 - x_3)(x_2 - x_1) > 0.$$

The positivity of (43.2) and (43.3) is a special case of the "rearrangement inequality." ∎

Suppose P is a partition of $\{1, \ldots, 8\}$ where one of the pairs is $\{k, 8\}$ with $1 < k < 8$. Then there is some other pair in the partition P of the form $\{1, j\}$ with j distinct from 1, k, and 8. The Lemma applies (with either $1 < j < k < 8$ or $1 < k < j < 8$) and $N_4(\{\{k, 8\}, \{1, j\}\}) < N_4(\{\{1, 8\}, \{k, j\}\})$. This shows that

$$N(P) = N\left(\{\{k, 8\}, \{1, j\}, \{x_3, y_3\}, \{x_4, y_4\}\}\right)$$
$$< N\left(\{\{1, 8\}, \{k, j\}, \{x_3, y_3\}, \{x_4, y_4\}\}\right),$$

so $N(P)$ is not a maximum. We can conclude that any partition maximizing N must include the pair $\{1, 8\}$.

Next, consider a partition P where one of the pairs is $\{1, 8\}$ and another is $\{k, 7\}$ with $2 < k < 7$. Then there is some other pair in the partition P of the form $\{2, j\}$ with j distinct from 1, 2, k, 7, and 8. The Lemma applies and $N_4(\{\{k, 7\}, \{2, j\}\}) < N_4(\{\{2, 7\}, \{k, j\}\})$. This shows that

$$N(P) = N\left(\{\{1, 8\}, \{k, 7\}, \{2, j\}, \{x_4, y_4\}\}\right)$$
$$< N\left(\{\{1, 8\}, \{2, 7\}, \{j, k\}, \{x_4, y_4\}\}\right),$$

so $N(P)$ is not a maximum. We can conclude that any partition maximizing N must include the pairs $\{1, 8\}$ and $\{2, 7\}$.

The only numbers remaining in any partition maximizing N are $\{3, 4, 5, 6\}$ and the Lemma applies one last time, so that P_2 is the unique partition maximizing N as claimed.

S2021-6A

(a) Let $a = 3$, $b = 4$, and $c = 7$.
(b) The left-hand side is equivalent to $109a + 91c - 89b$, which must equal 707 by the conditions of the problem. Rewriting, $2(10a + b) = 707 - 89(a + c - b)$. The left-hand side is even, which means $n = a + c - b$ must be odd. It is easy to see that $n < 8$, since $707 - 89 * 8 < 0$. Since a and b are single digits, the largest that $(10a + b)$ can be is 99, which means only $n = 7$ is a possibility. This

makes $(10a + b) = 42$, making $a = 4$ and $b = 2$, and hence $c = 5$. However, this makes $109a + 91c - 89b = 713$, not 707 as required.

S2021-6B Denoting the total by T and expanding give

$$T = \underline{abc} + \underline{cab} - \underline{bca} = 109a - 89b + 91c = 101(a - b + c) + (8a + 12b - 10c).$$

The second term is even and satisfies

$$-90 \leq 8a + 12b - 10c \leq 180,$$

so

$$T - 180 \leq 101(a - b + c) \leq T + 90. \tag{43.4}$$

For part (a), $T = 608$ is even, and the only even solution of (43.4) is $a - b + c = 6$, so $8a + 12b - 10c = 2$. For part (b), $T = 707$ is odd, and the only odd solution of (43.4) is $a - b + c = 7$, so $8a + 12b - 10c = 0$. These linear systems of equations can be solved simultaneously by reducing the doubly augmented matrix

$$\begin{bmatrix} 1 & -1 & 1 & | & 6 & 7 \\ 8 & 12 & -10 & | & 2 & 0 \end{bmatrix} \rightarrow \begin{bmatrix} 1 & 0 & 0.1 & | & 3.7 & 4.2 \\ 0 & 1 & -0.9 & | & -2.3 & -2.8 \end{bmatrix}.$$

For part (a), the equations reduce to $10a + c = 37$ and $10b - 9c = -23$, with the unique nonnegative single-digit solution vectors $a = 3$, $b = 4$, $c = 7$. For part (b), the equations reduce to $10a + c = 42$ and $10b - 9c = -28$, where $a = 4$ and $c = 2$ satisfy the first equation, but then $10b - 18 = -28$ has no nonnegative solution.

S2021-7 Let $p(x) = \sum_{k=0}^{n} a_k x^k$ with $a_k \geq 0$. Because $11^4 = 14641 > 2021$, the condition $p(11) = 2021$ implies p is of the form $a_3 x^3 + a_2 x^2 + a_1 x + a_0$, and because $2 \cdot 11^3 = 2662 > 2021$, $0 \leq a_3 \leq 1$. The condition that $p(1) = 21$ means the coefficients must sum to 21, and hence each of the coefficients cannot exceed 21. Considering the equation $a_3 11^3 + a_2 11^2 + a_1 11 + a_0 = 2021$ modulo 11 gives $a_0 \equiv 8 \pmod{11}$, so $a_0 = 8$ or $a_0 = 19$.

In the case $a_0 = 19$, the remaining coefficients must add to 2, but none of the triples $(a_3, a_2, a_1) = (0, 2, 0), (0, 1, 1), (0, 0, 2), (1, 1, 0),$ or $(1, 0, 1)$ satisfies $p(11) = 2021$.

In the case $a_0 = 8$ and $a_3 = 0$, $p(1) = 21$ implies $a_2 + a_1 = 13$ and $p(11) = 2021$ implies $121 a_2 + 11 a_1 = 2013$. The only solution of these two equations is $a_2 = 17$, $a_1 = -4$, contradicting $a_1 \geq 0$.

The only remaining case is $a_0 = 8$ and $a_3 = 1$. Then $p(1) = 21$ implies $a_2 + a_1 = 12$ and $p(11) = 2021$ implies $121a_2 + 11a_1 = 682$. The only solution of these two equations is $a_2 = 5$, $a_1 = 7$.

The conclusion is that $p(x) = x^3 + 5x^2 + 7x + 8$ and $p(10) = 1578$.

Remark from the Editors If one additionally assumes that each coefficient is a single digit ($0 \leq a_k \leq 9$), then the condition that $p(11) = 2021$ means that 2021 is the base-10 value of a number with most significant digit a_n and one's digit a_0, expressed in base-11. The value $p(10) = 1578$ can then be interpreted as the base-11 representation of 2021.

S2021-8 The number 288 has a unique factorization in exactly 10 numbers from the set $\{1, 2, 3, 4, 5\}$, $288 = 1^4 2^3 3^2 4^1 5^0$. We can fill in the different cells as follows:

- Because 288 cannot have any factors of five, we can fill in cells $(1, 1)$, $(5, 2)$, $(3, 3)$, $(2, 4)$, and $(4, 5)$ with fives.
- Because the cage of 288 must have 4 ones as factors, we can fill in cells $(1, 5)$, $(2, 2)$, $(4, 3)$, and $(2, 4)$, and $(3, 4)$ with ones. Thus, cell $(5, 1)$ must be a one, too.
- To get 3 factors of two in the 288 cage, cells $(1, 4)$ and $(1, 2)$ cannot be filled with a two. It follows that cell $(1, 3)$ must be a two, and from there it follows that cells $(3, 2)$ and $(4, 4)$ are also twos.
- The 288 cage is completed by inserting a four in cell $(1, 2)$ and threes in cells $(4, 2)$ and $(1, 4)$.
- The remainder of the cells are forced, and it can be verified that the other cage's clue is also satisfied.

The solution is as follows.

5	4	2	3	1
2	1	4	5	3
3	2	5	1	4
4	3	1	2	5
1	5	3	4	2

Chapter 44
2022 Solutions

The 56th competition was held at Indiana Wesleyan University, after 2 years without gathering in-person. Because of the transition back from remote to centralized formats, there were fewer teams than normal participating, and the meeting itself was held on a single day. While the competition was an in-person event, the scoring and reporting was done virtually. Results were shared publicly within a few days after the competition.

The 2022 problem set was prepared by Professor Colin McKinney at Wabash College.

The **2022 problem statements** begin on page 59 in Chap. 22.

S2022-1 This is a four-line locus problem, and so we should expect a conic section as the resulting plane curve. The required equation is

$$|x+3||1-x| = |y+4||2-y|.$$

There are two solutions, depending on the signs chosen when solving this equation with absolute values. One is the hyperbola

$$(y+1)^2 - (x+1)^2 = 5,$$

and the other is the circle

$$(x+1)^2 + (y+1)^2 = 13.$$

Historical note from the problem author Compare Apollonius' *Conics* III.54 [A] and Descartes' *La Géométrie* [De].

S2022-2 By the product rule, we need f and g to satisfy

$$f'g' = f'g + g'f.$$

For the sake of simplicity, choose $f(x) = x$ so that $f'(x) = 1$. Thus we require

$$g' = g + xg'.$$

Hence

$$\frac{g'}{g} = \frac{1}{1-x}.$$

Integrating on some interval where $g \neq 0$, we see that

$$\ln|g(x)| = -\ln|1-x| + C,$$

so that $g(x) = \dfrac{A}{1-x}$ for some constant A. Choosing $A = 1$ just to find a specific, a nonconstant example gives us functions $f(x) = x$ and $g(x) = \dfrac{1}{1-x}$. One can easily check that these functions satisfy the requirements. Another pair of functions that works is e^{2x} and e^{2x}.

Remark from the problem author This problem was inspired by ICMC **P1966-4** [AFMC], which asks a similar question but with quotients.

S2022-3

(a) Standard differentiation techniques yield

$$f'(x) = \frac{e^{-1/x}}{\left(1 - e^{-1/x}\right)^2 x^2}.$$

(b) Observe that, on $(0, \infty)$, f is an antiderivative of the integrand, by part (a), and both f and f' extend continuously to $[0, \infty)$. Therefore, by the Fundamental Theorem of Calculus,

$$\int_0^1 \frac{e^{-1/x}}{x^2(1 - e^{-1/x})^2}\,dx = \lim_{a \to 0^+} (f(1) - f(a))$$

$$= \frac{1}{e-1}.$$

Remark from the problem author This problem is a simplification of ICMC **P1971-3** [AFMC].

S2022-4 Let $B = I + A + A^2 + \cdots + A^{m-1}$. Observe that $B(I - A) = I - A^m = 0$. Since $A \neq I$, there is a vector $\vec{v}$ such that $A\vec{v} \neq \vec{v}$. Let $\vec{w} = (I - A)\vec{v} = \vec{v} - A\vec{v}$. Then $B\vec{w} = B(I - A)\vec{v} = 0$. Thus, $\vec{w}$ is a nonzero vector in the null space of B, and so $\det B = 0$.

Remark from the problem author This is the same as ICMC **P1999-8** [AFMC]. This new solution was contributed by Bob Foote.

S2022-5

(a) $n?$ is a nonstandard notation for the nth triangular number, given by

$$n? = T_n = \frac{(n)(n+1)}{2}.$$

Thus $2022? = 1011 \cdot 2023$, which clearly has no zeros at the end of its decimal representation.

(b) Divide 2022 by powers of 5 that are less than 2022, yielding $404 + 80 + 16 + 3 = 503$. Thus, there are 503 zeros at the end of the decimal representation.

Remark from the Editors Part (b) is similar to ICMC **P2012-2**—see the solution **S2012-2** in Chap. 35.

S2022-6 No. Let G be the group, and suppose by way of contradiction that H and K are proper subgroups with $G = H \cup K$. Because $H \not\subseteq K$, we can take an element $h \in H \setminus K$. Similarly, let $k \in K \setminus H$, and consider the element $hk \in G$. Then $hk \in H$ or $hk \in K$. If $hk \in H$, then $h^{-1}hk \in H \implies k \in H$, and if $hk \in K$, then $hkk^{-1} \in K \implies h \in K$; either case leads to a contradiction.

Remark from the problem author This is the same as ICMC **P1972-7** [AFMC].

S2022-7 Note that $g(x) = g(y) \implies g^m(x) = g^m(y) \implies x = y$. Therefore g is injective. By a well-known result from real analysis (not proved here, but which can be proved using the Intermediate Value Theorem), any injective real-valued continuous function on an interval must be strictly monotone on the interval. If we suppose that g is increasing on $[0, 1]$, then for $x \in [0, 1]$,

$$x > g(x) \implies g(x) > g^2(x) \implies \ldots \implies g^{m-1}(x) > g^m(x),$$

hence $x > g^m(x) = x$, a contradiction. Similarly, if $x < g(x)$, we have $x < g^m(x) = x$, another contradiction. Hence $g(x) = x$. If on the other hand g is decreasing, then g^2 is increasing, and the above argument applies to show $g^2(x) = x$. Hence, in either case, $g^2(x) = x$.

Remark from the problem author This problem and solution were both adapted from ICMC **P1983-9** [AFMC]. See also ICMC **P1977-1**.

S2022-8 The following example is constructed using the increasing sequence of triangular numbers:

$$(0, 1, 3, 6, 10, \ldots, T_a, \ldots),$$

defined by $T_a = \frac{1}{2}a(a+1)$ or recursively by $T_0 = 0$ and $T_{a+1} = T_a + a + 1$. Define $f : \mathbb{N} \times \mathbb{N} \to \mathbb{N}$ by

$$f(m, n) = m + T_{m+n-2}. \tag{44.1}$$

To show f is surjective, consider $x \geq 1$, and let $a = \max\{y \in \mathbb{Z} : T_y < x\}$, so that $T_a < x \leq T_{a+1}$. Then $0 < x - T_a \leq T_{a+1} - T_a = a + 1$, and one can verify that

$$f(x - T_a, a + 2 - (x - T_a)) = x.$$

To show f is injective, suppose there are two input pairs; label the pair with the smaller or equal sum of components by (m, n) and the other pair by (m^*, n^*), so $m + n \leq m^* + n^*$.

We want to show that assuming the outputs are equal,

$$f(m, n) = f(m^*, n^*)$$
$$m + T_{m+n-2} = m^* + T_{m^*+n^*-2} \tag{44.2}$$

leads to the conclusion $m = m^*$ and $n = n^*$.

Case 1. $m + n < m^* + n^*$. Then, using the increasing property of the T_a sequence,

$$0 = T_{m^*+n^*-2} - T_{m+n-2} + m^* - m$$
$$\geq T_{m+n-1} - T_{m+n-2} + m^* - m$$
$$= m + n - 1 + m^* - m$$
$$= m^* + n - 1$$
$$\geq 1,$$

which is a contradiction.

Case 2. $m+n = m^*+n^*$. It follows immediately from (44.2) that $m = m^*$; thus, in this case, $n = n^*$.

Historical note from the problem author The $f(m, n)$ formula (44.1) was given by Cantor, see [C].

Remark from the Editors Although there may be many formulas for bijections $\mathbb{N} \times \mathbb{N} \to \mathbb{N}$, the Fueter–Pólya Theorem states that $f(m, n)$ and the expression $f(n, m) = n + T_{m+n-2}$ are the only quadratic polynomials defining such a bijection. The appearance of Cantor's construction in problems for students dates back to [M].

Chapter 45
2023 Solutions

The 57th competition was held at Indiana University Kokomo for the first time.

The 2023 problem set was prepared by Professor Emeritus Robert Foote at Wabash College.

The **2023 problem statements** begin on page 61 in Chap. 23.

The 2023 competition's tied-for-first place team from Rose-Hulman. Left to Right: Professor Josh Holden, Nathan Chen, Connor Lane, and Alexa Renner.

S2023-1 We seek the maximum and minimum speeds of the particle, where speed is the magnitude (absolute value) of velocity. Let

$$v(t) = x'(t) = 7 + 6t - t^2.$$

Looking for the max/min of $v(t)$ on the $[0, 10]$ interval, we have that $v'(t) = 0$ only at $t = 3$. Check that

$$v(0) = 7, \qquad v(3) = 16 \quad \text{and} \quad v(10) = -33.$$

The fastest it goes is 33 ft/s when $t = 10$. Since $v(t)$ changes sign, the slowest it goes is 0 ft/s, which happens when $t = 7$.

S2023-2 Consider

$$\sin^2 70° = \sin^2 50° + \sin^2 60° - 2 \sin 50° \sin 60° \cos 70°.$$

This is the Law of Cosines for a triangle with sides $\sin 50°$, $\sin 60°$, and $\sin 70°$, assuming such a triangle exists. Since $50° + 60° + 70° = 180°$, there are triangles with angles $50°$, $60°$, and $70°$. The Law of Sines states that

$$\frac{\sin 50°}{a} = \frac{\sin 60°}{b} = \frac{\sin 70°}{c},$$

where a, b, and c are the respective side lengths. This common ratio can be any positive number—it simply scales the size of the triangle. When the ratio is 1, we have that

$$a = \sin 50°, \quad b = \sin 60°, \quad \text{and} \quad c = \sin 70°$$

are the sides of a triangle.

S2023-3 Let $M_X = \{(x, y, z) \in [0, 1]^3 \mid x \geq y \text{ and } x \geq z\}$. This is the portion of the cube on which x is the maximum. Then

$$\int_0^1 \int_0^1 \int_0^1 f(x, y, z) \, dx \, dy \, dz = 3 \iiint_{M_X} x \, dx \, dy \, dz$$

$$= 3 \int_0^1 \int_0^x \int_0^x x \, dy \, dz \, dx = 3 \int_0^1 x^3 \, dx = 3/4.$$

Remark from the problem author This integral has a nice interpretation in probability. It is the expected value of $\max\{x, y, z\}$ when x, y, and z are taken to be independent, random variables on $[0, 1]$ with a uniform distribution.

S2023-4A Differentiating both sides yields this differential equation:

$$f(x) + f(-x) = f'(x) + f'(-x).$$

The next step is not forced on us, but remembering that we are looking for *some* function that works, it seems reasonable to consider functions that satisfy $f'(t) = f(t)$, namely $f(t) = ae^t$. It is easy (and necessary) to check that this satisfies the integral equation. The desired function of this form is $f(x) = 2023e^t$.

S2023-4B The symmetries of the equation around $x = 0$ may lead you to think about even and odd functions. If so, you will quickly notice that the left side of the equation is 0 if f is even, and the right side is 0 if f is odd. Thus, if a function satisfies the equation, there must be some relation between its even and odd parts. The simplest function that has both even and odd parts is $f(x) = ax + b$. If you plug this into both sides of the equation, it is a solution if and only if $a = b$, so the desired function of this form is $f(x) = 2023x + 2023$.

S2023-4C More generally, any function $f(x)$ is the sum of its even and odd parts, $f_e(x) = \frac{1}{2}(f(x) + f(-x))$ and $f_o(x) = \frac{1}{2}(f(x) - f(-x))$. Then, assuming only that f is integrable and f_e is continuous, the integral equation implies that f_o is differentiable, and $f'_o(x) = f_e(x)$. This leads to some unusual functions satisfying the equation, for example, choosing $f_e = 2|x| + 2023$ gives a nondifferentiable solution $f(x) = x|x| + 2|x| + 2023x + 2023$.

S2023-5A Proof by induction. Let $g_n = f_{n+1}f_{n-1} - f_n^2$. We have $g_2 = 2 \cdot 1 - 1 = (-1)^2$ for the base case. Suppose $g_{n-1} = (-1)^{n-1}$. Then

$$g_n = f_{n+1}f_{n-1} - f_n^2 = (f_n + f_{n-1})f_{n-1} - (f_{n-1} + f_{n-2})f_n$$
$$= f_{n-1}^2 - f_n f_{n-2} = -g_{n-1} = (-1)^n.$$

S2023-5B Another proof also uses induction and a nice matrix equation that makes the algebra a bit cleaner. Let $F_n = \begin{bmatrix} f_{n+1} & f_n \\ f_n & f_{n-1} \end{bmatrix}$. We need to show $\det F_n = (-1)^n$ for $n \geq 2$, where the base case $n = 2$ is as in the previous solution. Let $E = \begin{bmatrix} 1 & 1 \\ 1 & 0 \end{bmatrix}$, and note that

$$F_n = EF_{n-1}. \tag{45.1}$$

Suppose, for $n > 2$, $\det F_{n-1} = (-1)^{n-1}$. Then

$$\det F_n = \det E \det F_{n-1} = (-1)(-1)^{n-1} = (-1)^n.$$

Remark from the Editors A matrix product like (45.1) was considered in **S2003-6** in Chap. 26.

S2023-6 Let L be the line parallel to $\overline{AB}$ that is also tangent to γ. Let Z be the intersection of L and $\overleftrightarrow{CH}$. Note that $ZX = HY$, since $\overline{ZX}$ and $\overline{HY}$ are symmetric across the diameter of γ parallel to L and $\overline{AB}$ (dotted line). Since Z is between C and X (see the next paragraph), we have $CX > ZX = HY$.

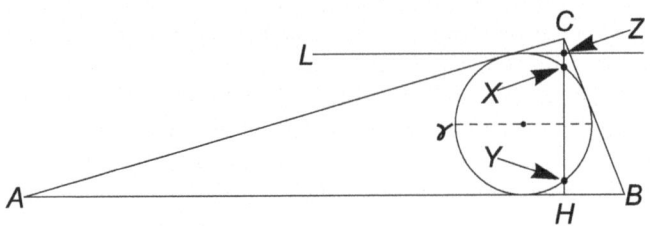

To see that Z is between C and X, note that C and H are on opposite sides of L, since L intersects the interior of $\triangle ABC$ but not $\overline{AB}$. On the other hand, X and H are on the same side of L, since γ is between L and $\overleftrightarrow{AB}$. Thus, C and X are on opposite sides of L.

S2023-7 First, for each face-down card, consider the probability that it is black. (This is a variation of the classic Monty Hall problem in which the black card is the prize. This is worth looking up if you are unfamiliar with it.) Let $P(c$ is B$)$ denote the probability that card c is black. Initially (while the cards are all face down), $P(c$ is B$) = 1/4$ for each c. Let W be the card you choose, let X be the card your friend turns up, and let Y and Z be the other two cards. When your friend turns up a card, it adds information about X, Y, and Z, but it adds no information about W. In particular, $P(W$ is B$)$ is still $1/4$ and $P(X$ is B$) = 0$. By symmetry, $P(Y$ is B$) = P(Z$ is B$)$. The four probabilities sum to 1, and so we have

$$P(W \text{ is B}) = 1/4, \quad P(Y \text{ is B}) = 3/8, \quad \text{and} \quad P(Z \text{ is B}) = 3/8.$$

Then (with R for red)

$$P(W \text{ is R}) = 3/4, \quad P(Y \text{ is R}) = 5/8, \quad \text{and} \quad P(Z \text{ is R}) = 5/8.$$

Remark from the problem author The claim that the revealed card adds no information about W can be explained by drawing a probability tree for the problem.

S2023-8 Consider two beads that are about to collide. One moves to the right and one to the left. After the collision, there is still one moving to the right and one to the left. Thus, the numbers of right-moving beads and left-moving beads are constant. After all the collisions there will still be n right-moving beads and k left-moving beads, which answers part (a). If we ignore the bead colors, assume they are negligibly small, and focus on the directions they move, the dynamics are the same if we allow the beads to pass through each other instead of collide and rebound. Then every right-moving bead passes through every left-moving bead. There are nk such pass-through events (possibly some at simultaneous times, but spatially separated because the right-moving beads maintain their initial pairwise distances from each other), so in the original collision model, there are nk collisions, which answers part (b).

Part III
More History of the ICMC

Chapter 46
Top Scoring Teams

The student participants in the 2015 ICMC, held at Taylor University

The following table contains the names of the winning teams for the period covered by this volume. It includes some ties and, notably, reflects some early years in which the section maintained the practice of not recognizing multiple high-scoring teams from the same school.

Year		School	Name
2001	1	Rose-Hulman Institute of Technology	Andrew Chi
			Ann Chi
	2	Taylor University	Shawn Alspaugh
			David Aukerman
			Noah DeLong
	3	Wabash College	Bogdan Ianev
			Nigel Nunoo
			Nick Roersma
	3	Indiana University	Greg Alexander
			Craig Citro
			Jason Grimblat
2002	1	IUPUI	Bobby Ramsey
			Muris Ridzal
			Jon Landy
	2	Wabash College	Shiv Karunakaran
			Bogdan Ianev
			Daniel Smith
	3	Taylor University	David Aukerman
			Shawn Alspaugh
			Noah DeLong
	3	Indiana University	Craig Citro
			Jason Grimblat
			Seth Quackenbush
2003	1	Indiana—Purdue Fort Wayne	Christian MacLeod
			Jeff Wilkins
			Kevin Chlebik
	2	IUPUI	Robert Walsman
			Scott Pollum
			Charles Tam
	3	Wabash College	Bogdan Ianev
			Daniel Smith
			Mick Roersma
2004	1	Rose-Hulman Institute of Technology	Guy Srinivasan
			Alex VanBrunt
			Kellan Wampler
	2	Taylor University	Michael Anderson
			Mike Assis
			Shawn Burford
	3	Ball State University	

(continued)

Year		School	Name
2005	1	Purdue University	Kyle Riggs
			Brad Rodgers
			Chris Willmore
	2	Rose-Hulman Institute of Technology	Stephen Dupal
			Guy Srinivasan
			Kellan Wampler
	3	Rose-Hulman Institute of Technology	Peter Fine
			Curtis Katinas
			Angela Smiley
2006	1	Rose-Hulman Institute of Technology	Stephen Dupal
			Robert Lemke-Oliver
			Hari Ravindran
	2	Purdue University	Kyle Riggs
			Brad Rodgers
			Chris Willmore
	3	Ball State University	
2007	1	Rose-Hulman Institute of Technology	Robert-Lemke-Oliver
			Hari Ravindran
			Amanda Rohde
	2	Ball State University	Neal Coleman
			Wayne Drake
			Nathan Pappas
	3	Purdue University	Jonathon Nistor
			Nate Oriow
			Prateek Tandon
	3	Taylor University	Jeremy Erickson
			David Kasper
			Matthew Russell
2008	1	Purdue University	Noah Blach
			Nate Oriow
			Brad Rodgers
	2	Rose-Hulman Institute of Technology	Robert Lemke-Oliver
			Hari Ravindran
	3	Taylor University	Jeremy Erickson
			Matthew Russell
			Jonathan Schrock
2009	1	Taylor University	Jeremy Erickson
			Matthew Russell
			Joseph Seaborn
	2	Indiana University	Yun William Yu
			Carlo Angiuli
			Peter Lunts

(continued)

Year		School	Name
	3	IUPUI	Jeffrey Shen
			Lyndon Ji
2010	1	Indiana University	Carlo Angiuli
			John Brown
			Miles Dillon Edwards
	2	Taylor University	Jacob Erb
			Matthew Russell
			Joe Seaborn
	3	Indiana Wesleyan University	Tyler Carrico
			Sam Johnston
			Luke Nelson
2011	1	Indiana University	John Brown
			Kevin Carlson
			Miles Dillon Edwards
	2	IUPUI	Kyle Firestone
			Youkow Homma
			William Karr
	3	Taylor University	David Ebert
			Jacob Erb
			Daniel Kasper
2012	1	Indiana University	John Brown
			Miles Dillon Edwards
			Timothy Zakian
	2	Wabash College	Yifei Sun
			Anh Tran
			Tyler Koch
	3	Indiana—Purdue Fort Wayne	Valentin Bogun
			Samuel Carolus
			Altun Shukurlu
2013	1	Valparaiso University	Timothy Goodrich
			Michael Stuck
			Ruyue (Julia) Yuan
	2	Wabash College	David Gunderman
			Jia Qi
			Xidian Sun
	3	Taylor University	Daniel Crane
			Ethan Gegner
			Josh Kiers
2014	1	Indiana University	Tom Dauer
			Jonathan Hawkins
			Max Zhou
	1	Taylor University	Ethan Gegner
			Josh Kiers
			Claire Spychalla

(continued)

Year		School	Name
	3	Indiana—Purdue Fort Wayne	Vreneli Brenneman
			Guchen Liu
			Altun Shukurlu
	3	Taylor University	Sam Judge
			Jordan Melendez
			Justin Wydra
2015	1	Taylor University	Daniel Crane
			Josh Kiers
			Claire Spychalla
	2	Rose-Hulman Institute of Technology	Ian Ludden
			Christian Schulz
			Lujia Zhang
	3	Indiana—Purdue Fort Wayne	Vreneli Brenneman
			Altun Shukurlu
			Sofia Sorokina
2016	1	Indiana University	Tom Dauer
			Baptiste Dejean
			Max Zhou
	2	Indiana Wesleyan University	Jeremy David
			Chriss Foss
			Luke Hayden
	3	Rose-Hulman Institute of Technology	Jacob Hiance
			Adam Michael
2017	1	Indiana University	Ben Briggs
			Andrew Henderson
			Baptiste Dejean
	2	Wabash College	Tu Nguyen
			Ngoc Tran
			Thang Nguyen
	3	Earlham College	Vuong Khuat
			Hiep Nguyen
			Son Tran
2018	1	Indiana University	Anthony Coniglio
			Baptiste Dejean
			Nathanial Lowry
	2	Purdue University	Israel Baker
			Kevin LaMaster
			Junyao Wu
	3	University of Notre Dame	Kyle Duffy
			Matt Schoenbauer
			Caitlyn Booms
	3	Indiana—Purdue Fort Wayne	Vincent Rivera
			Giang Le
			Duy Anh Do

(continued)

Year		School	Name
2019	1	Rose-Hulman Institute of Technology	Zixin Fan
			Ruizin Feng
			Jiashen Hu
	2	Indiana University	Anthony Coniglio
			Mathanial Lowry
			Tiancheng Xu
	3	Taylor University	Drew Anderson
			Jordan Crawford
			Alexander McFarland
2021	1	Purdue University	Ethan Brady
			Sterling Saint-Rain
			Chiara Travesset
	2	Rose-Hulman Institute of Technology	Reed Phillips
			Bohdan Vakhitov
			Max Li
	3	Valparaiso University	Eric Burkholder
			Spencer Gannon
			Marcus Engstrom
	3	Purdue University	Max Martinez
			Kyle Kraft
			Colton Griffith
2022	1	Rose-Hulman Institute of Technology	Reed Phillips
			Ben Lyons
	2	Purdue University	Kyle Kraft
			Max Martinez
			Colton Griffen
	2	IUPUI	Sara Hiatt
			Croix Gyurek
			Clayton Kelley
2023	1	Notre Dame	Samuel Dekleva
			Molly MacDonald
			Philip Templeman
	1	Rose-Hulman Institute of Technology	Nathan Chen
			Connor Lane
			Alexa Renner
	3	Notre Dame	Andrew Brooks
			Gavin Dooley
			Zachary Joseph

Chapter 47
Updates to *A Friendly Mathematics Competition*

Some Errata for the 1966–2000 Problem Book

Page numbers for these corrections refer to the publication [AFMC].

- p. 15: The team member's name is Tom Sellke; he also appears on p. 16.
- p. 30: The word in line 7 of **P1987-3** should be "falling."
- p. 30: The distance formula in **P1987-5** should be

$$|XY| = \left((y_1 - x_1)^2 + (y_2 - x_2)^2 + (y_3 - x_3)^2 + (y_4 - x_4)^2\right)^{1/2}.$$

- p. 33: The notation should indicate an infinite sequence

$$\{x_1, x_2, \ldots, x_k, \ldots\}.$$

- p. 38: The team member's name is Mike Bolt.
- p. 42: The symbol in **P1996-5** should be "degrees": $\angle A = 90°$.
- p. 60: The symbol in paragraph 2, line 2, of **S1967-8** should be a bold **Q**.
- p. 62: The symbol in line 4 of **S1968-3** should have a prime for the derivative: $tf'(tx)$. See also comments in section "New Solutions for Old Problems".
- p. 87: The expression in line 2 of **S1976-4** should have an equal sign: $f'(1) = 3\ln 3 - 4 < 3.3 - 4 < 0$.
- p. 88: The symbol in the last line of **S1976-6** should be a strict inequality: $f(x) < 0$.
- p. 126: The number in line 3 of **S1987-6** should be $2\pi(4000)/4$.
- p. 131: The equation in **S1989-3** should be $2(f + c) = a + d + g$.
- p. 167: The symbol in **S1998-4** in the last sentence on this page should be a bold **F**.

- p. 172: In **S1999-3**, some y terms should be y^2, so that line (ii) is

$$5x^2 + 2xy + y^2 \leq 2x^2 + 2\sqrt{2}x\sqrt{x^2 + y^2} + x^2 + y^2$$

and line (iv) is

$$x^2 + 2xy + y^2 \leq 2(x^2 + y^2).$$

New Solutions for Old Problems

P1968-3 Let $f : \mathbb{R}^n \to \mathbb{R}^n$ be a differentiable function such that $f(tx) = tf(x)$ for $x \in \mathbb{R}^n$ and $t > 0$. Show that f is linear.

The solution appearing in [AFMC] is short and elementary but seems to use an unstated, and unnecessary, assumption that f is continuously differentiable.

Here we state and prove a more general result.

Theorem Let $f : \mathbb{R}^n \to \mathbb{R}^m$ be a function which is differentiable at the origin, and such that $f(t\vec{x}) = tf(\vec{x})$ for $\vec{x} \in \mathbb{R}^n$ and $t > 0$. Then f is linear.

Proof By the definition of multivariable differentiability, there is some constant $m \times n$ matrix M so that

$$\lim_{\vec{h} \to \vec{0}} \frac{f(\vec{0} + \vec{h}) - f(\vec{0}) - M \cdot \vec{h}}{\|\vec{h}\|} = \vec{0}.$$

We also have, by hypothesis,

$$f(\vec{0}) = f(2 \cdot \vec{0}) = 2 \cdot f(\vec{0}) \implies f(\vec{0}) = \vec{0}.$$

So, the differentiability property is

$$\lim_{\vec{h} \to \vec{0}} \frac{f(\vec{h}) - M \cdot \vec{h}}{\|\vec{h}\|} = \vec{0}.$$

Fix $\vec{x} \neq \vec{0}$, and let ϵ be an arbitrary positive number. Considering $\frac{\epsilon}{\|\vec{x}\|} > 0$, there is a corresponding $\delta > 0$ so that if $0 < \|\vec{h}\| < \delta$, then

$$\frac{\|f(\vec{h}) - M \cdot \vec{h}\|}{\|\vec{h}\|} < \frac{\epsilon}{\|\vec{x}\|}.$$

Pick $t > 0$ so that $\|t\vec{x}\| < \delta$. Then

$$\frac{\|f(t\vec{x}) - M \cdot (t\vec{x})\|}{\|t\vec{x}\|} < \frac{\epsilon}{\|\vec{x}\|}$$

$$\implies \frac{1}{t}\|tf(\vec{x}) - tM \cdot \vec{x}\| < \epsilon$$

$$\implies \|f(\vec{x}) - M \cdot \vec{x}\| < \epsilon,$$

which, since ϵ and $\vec{x}$ were arbitrary, implies $f(\vec{x}) = M \cdot \vec{x}$ for any $\vec{x} \neq \vec{0}$. The conclusion is that $f(\vec{x})$ is equal to the linear expression $M \cdot \vec{x}$ for all $\vec{x} \in \mathbb{R}^n$. ∎

P1969-6 Assume that f has a continuous second derivative, that $a < b$, that $f(a) = f(b) = 0$, and that $|f''(x)| \leq M$ on $a \leq x \leq b$. Prove that

$$\left|\int_a^b f(x)dx\right| \leq \frac{M}{12}(b-a)^3.$$

The solution appearing in [AFMC] identifies this inequality as a special case of the error estimate for the Trapezoidal Rule for integrals and then refers the reader to the literature to find a proof. Here we sketch such a proof for the claimed special case.

For t in an open interval containing $[a, b]$, define this difference of areas:

$$g(t) = \int_a^t f(x)dx - \frac{1}{2}(t-a)f(t),$$

so that $g(a) = 0$ and $g(b) = \int_a^b f(x)dx$ (using $f(b) = 0$). g is differentiable, with

$$g'(t) = \frac{1}{2}f(t) - \frac{1}{2}(t-a)f'(t),$$

and $g'(a) = 0$ (using $f(a) = 0$). The second derivative is continuous by hypothesis

$$g''(t) = -\frac{1}{2}(t-a)f''(t).$$

Using the Fundamental Theorem of Calculus,

$$\left|\int_a^b f(x)dx\right| = |g(b) - g(a)| = \left|\int_a^b g'(x)dx\right|$$

$$= \left|\int_a^b \left(g'(a) + \int_a^x g''(y)dy\right)dx\right|$$

$$\leq \int_a^b \int_a^x |g''(y)| \, dy \, dx = \int_a^b \int_a^x \left| -\frac{1}{2}(y-a) f''(y) \right| dy \, dx$$

$$\leq \int_a^b \int_a^x \frac{1}{2}(y-a) M \, dy \, dx = \frac{M}{4} \int_a^b (x-a)^2 \, dx = \frac{M}{12}(b-a)^3.$$

P1982-5 A real-valued function f of a real variable is said to satisfy a Hölder condition with exponent α if there is a constant c such that $|f(x) - f(y)| \leq c|x-y|^\alpha$ for all x, y. Wherever these functions are used, α is restricted to be ≤ 1. Can you explain why?

The solution appearing in [AFMC] states that if $\alpha \leq 1$, then f is uniformly continuous on its domain. This is true for f on an interval where the condition is satisfied with $0 < \alpha \leq 1$, for example, $f(x) = |x|^\alpha$ on the interval $[-1, 1]$, but it seems to be an incomplete answer to the question.

Here we state and prove the following result.

Theorem If $f(x)$ has the property that there exist $c \geq 0$ and $\alpha > 1$ such that $|f(x) - f(y)| \leq c|x - y|^\alpha$ for all x, y in an open interval (a, b), then f is constant on (a, b).

Proof Fix $y \in (a, b)$, and consider this expression depending on $x \in (a, b), x \neq y$:

$$\left| \frac{f(x) - f(y)}{x - y} \right| = \frac{|f(x) - f(y)|}{|x - y|}$$

$$\leq \frac{c|x - y|^\alpha}{|x - y|} = c|x - y|^{\alpha - 1}.$$

Because $\alpha - 1 > 0$, the Squeeze Theorem for limits applies to show that the limit exists:

$$f'(y) = \lim_{x \to y} \frac{f(x) - f(y)}{x - y} = \lim_{x \to y} c|x - y|^{\alpha - 1} = 0.$$

Since the derivative is zero at every point y in the interval, the function is constant (as in **P2016-6**). ∎

P1986-3 Evaluate i^i.

The solution appearing in [AFMC] gives a correct answer, $i^i = e^{-\pi/2}$. However, complex exponentiation is multivalued, so there are infinitely many other answers (which in this case are all real numbers): For any integer n,

$$i^i = (e^{(\pi i/2) + 2\pi i n})^i = e^{-\pi/2 - 2\pi n}.$$

This sequence can also be explained in terms of the multivalued property of the complex logarithm:

$$i^i = e^{i \log(i)} = e^{i((\pi i/2)+2\pi i n)}.$$

P1992-6c For an arbitrary 2×2 matrix A, what is the maximum number of integral matrices B that can satisfy $B^2 = A$?

The solution appearing in [AFMC] gives the example $A = \begin{bmatrix} 2^n & 0 \\ 0 & 2^n \end{bmatrix}$ and the $2(n+1)$ matrices $B = \pm \begin{bmatrix} 0 & 2^i \\ 2^{n-i} & 0 \end{bmatrix}$, for $0 \le i \le n$, concluding that there is no maximum. Another example, where there are infinitely many solutions, is $A = \begin{bmatrix} 0 & 0 \\ 0 & 0 \end{bmatrix}$ and the matrices $B = \begin{bmatrix} 0 & i \\ 0 & 0 \end{bmatrix}$, for any integer i.

Appendix

Location Index

The following educational institutions in Indiana appear in this book (or the earlier volume [AFMC], for 1966–2000) as host venues for Indiana MAA meetings, or as the home field for winning teams or problem writers.

1. Anderson University
2. Ball State University
3. Butler University
4. DePauw University
5. Earlham College
6. Franklin College
7. Goshen College
8. Hanover College
9. Huntington University
10. Indiana State University
11. Indiana University Bloomington
12. Indiana University East
13. Indiana University Kokomo
14. Indiana University Northwest
15. Indiana University - Purdue University Fort Wayne
16. Indiana University - Purdue University Indianapolis
17. Indiana Wesleyan University
18. Manchester College
19. Marian College
20. Purdue University Calumet
21. Purdue University North Central
22. Purdue University West Lafayette
23. Rose-Hulman Institute of Technology
24. St. Joseph's College
25. Saint Mary-of-the-Woods College
26. Saint Mary's College
27. Taylor University
28. Trine University (formerly Tri-State University)
29. University of Evansville
30. University of Indianapolis
31. University of Notre Dame
32. University of Southern Indiana
33. Valparaiso University
34. Wabash College

Appendix 215

References and Photo Credits

[A] Apollonius, *On Conic Sections*, translated by R. Catesby Taliaferro. University of Chicago Great Books 11 (Encyclopædia Britannica, Inc., 1952)

[C] G. Cantor, *Contributions to the Founding of the Theory of Transfinite Numbers*, translated by P. Jourdain (Dover Publications, Inc., 1955). Reprint of 1915 publication by Open Court Pub. Co., Chicago

[CZ] A. Coffman, Y. Zhang, Vector fields with continuous curl but discontinuous partial derivatives. Am. Math. Mon. (10) **127**, 911–917 (2020)

[Da] H. Davis, The first meeting of the Indiana Section. Am. Math. Mon. (1) **32**, 37 (1925)

[De] R. Descartes, *The Geometry*, translated by D. Smith, M. Latham, with a facsimile of the first edition, 1637 (Dover Publications, Inc., 1954), reprint of 1925 publication by Open Court Pub. Co., Chicago

[ME_{22}] Margaret Edson, private communication with RG, November 2022

[ME_{23}] Margaret Edson, private communication with RG, June 2023

[MA] Mary (Edson) Ales, private communication with RG, April 2023

[GKL] G. Gilbert, M. Krusemeyer, L. Larson, *The Wohascum County Problem Book*. Dolciani Mathematical Expositions 14 (Mathematical Association of America, 1993)

[AFMC] R. Gillman (ed.), *A Friendly Mathematics Competition. 35 Years of Teamwork in Indiana*. MAA Problem Books Series 8 (MAA Press: An Imprint of the American Mathematical Society, 2003)

[G] H. Grossman, The twelve-coin problem. Scr. Math. **11**, 360–361 (1945)

[GN] R. Guy, R. Nowakowski, Coin-weighing problems. Am. Math. Mon. (2) **102**, 164–167 (1995)

[HWr] G. Hardy, E. Wright, *An Introduction to the Theory of Numbers*, 5th edn. (Oxford, 1979)

[HWi] K. Hardy, K. Williams, *The Green Book of Mathematical Problems* (Dover Publications, Inc., 1997). Reprint of 1985 publication by Integer Press, Ottawa

[K_1] J. Kürschák, G. Hajós, G. Neukomm, J. Surányi (eds.), *Hungarian Problem Book I. Based on the Eötvös Competitions*, 1894 – 1905, translated by E. Rapaport. New Mathematical Library 11 (Random House, 1963)

[K_2] J. Kürschák, G. Hajós, G. Neukomm, J. Surányi (eds.), *Hungarian Problem Book II. Based on the Eötvös Competitions*, 1906 – 1928, translated by E. Rapaport. New Mathematical Library 12 (Random House, 1963)

[M] G. Mathews, *Theory of Numbers. Part I* (Deighton, Bell & Co., Cambridge, 1892)

[M_1] P. Mielke, Improper Integrals in Abstract Spaces, Purdue University Ph.D. dissertation, 1951

[M_2] P. Mielke, Rational points on the number line. Math. Teach. (6) **63**, 475–479 (1970)

[M_3] P. Mielke, A tiling of the plane with triangles. Two-Year Coll. Math. J. (5) **14**, 377–381 (1983)

[PFW] Photograph courtesy J. Whitcraft, Purdue University Fort Wayne Communications and Marketing

[RL] R. Rusczyk, S. Lehoczky, *The Art of Problem Solving. Volume 2: and Beyond*, 7th edition with separate Solutions Manual (AoPS Inc., 2013)

[S] D. Struik (ed.), *A Source Book in Mathematics*, 1200 – 1800 (Harvard University Press, 1969)

[TU] Photograph courtesy J. Case, Taylor University Department of Mathematics

[WCLib] Photographs of Paul Mielke, P-21, Wabash Photograph Collection, Robert T. Ramsay Jr. Archival Center, Lilly Library, Wabash College, Crawfordsville

[WCNews] P. Mielke, Jr. et al., Memorial Service for Professor Emeritus Paul Mielke '42, Wabash College web site news page (2008). https://www.wabash.edu/news/story/6013

[W] S. Weintraub, An observation on average velocity. Am. Math. Mon. (4) **130**, 384 (2023)

[WHS] Peter Edson Papers, Archives of the Wisconsin Historical Society, Madison. http://digital.library.wisc.edu/1711.dl/wiarchives.uw-whs-mss00730

Index

A
Algebra
  abstract
    P2003-1, 7
    P2005-4, 11
    P2008-1, 17
    P2015-7, 37
    P2016-2, 39
  groups
    P2002-5, 6
    P2007-6, 16
    P2009-3, 19
    P2012-5, 27
    P2013-4, 29
    P2013-5, 29
    P2014-7, 32
    P2016-7, 40
    P2019-7, 52
    P2022-6, 60
  matrix
    P1992-6c, 211
    P2003-4, 7
    P2003-6, 8
    P2004-3, 9
    P2005-5, 11
    P2007-5, 16
    P2008-7, 18
    P2010-4, 21
    P2011-8, 25
    P2012-4, 27
    P2014-3, 31
    P2016-5, 40
    P2017-6, 47
    P2018-4, 49

    P2019-5, 52
    P2021-1, 57
    P2022-4, 59
    S2013-6B, 134
    S2021-6B, 185
    S2023-5B, 195
  polynomials
    P2007-1, 15
    P2012-3, 27
    P2021-2, 57
    P2021-7, 58
    S2008-7A, 102
    S2009-4, 104
    S2016-5, 154
  rings
    P2002-5, 6
    P2006-5, 14
    S2005-5A, 87

B
Binomial Theorem
  S2005-8, 88
  S2012-1, 121
  S2014-8, 142
  S2015-2A, 146

C
Calculus
  differentiation
    P2015-8, 37
    P2016-6, 40
    P2021-2, 57
    P2023-1, 61
    S2009-6B, 105

Calculus (*cont.*)
  first semester
    P2001-2, 3
    P2004-6, 10
    P2006-1, 13
    P2006-4, 13
  infinite series
    P2001-6, 3
    P2002-2, 5
    P2003-5, 8
    P2004-2, 9
    P2006-2, 13
    P2009-2, 19
    P2017-2, 46
    S2010-5, 110
  integration
    P1969-6, 209
    P2005-1, 11
    P2007-3, 15
    P2012-7, 27
    P2022-3, 59
    P2023-4, 61
    S2006-3, 92
  limits
    P2005-6, 11
    P2007-3, 15
    P2010-5, 22
    P2011-3, 24
    P2011-6, 25
    P2012-7, 27
    P2014-1, 31
    P2014-2, 31
  multivariable
    P1968-3, 208
    P2008-4, 17
    P2011-3, 24
    P2012-8, 28
    P2023-3, 61
    S2019-8, 179
  sequences
    P2004-4, 9
    P2005-3, 11
    P2008-5, 17
    P2013-1, 29
    P2017-5, 47
    P2018-5, 49
    P2019-6, 52
  volumes
    P2008-4, 17
    P2011-2, 23
    P2019-8, 53
    S2014-3A, 138
    S2015-3, 147
Cauchy's Theorem
  S2015-7, 149
Cayley graph
  S2019-7, 178
Cayley-Hamilton Theorem
  S2008-7A, 102
Combinatorics
  P2001-3, 3
  P2002-4, 5
  P2010-1, 21
  P2013-3, 29
  P2015-2, 35
  P2016-8, 41
  P2017-7, 47
  P2018-8, 50
  P2019-3, 51
  P2021-5, 57
  P2023-8, 62
Comparison Test
  S2001-6, 68
  S2002-2, 72
  S2003-5, 78
  S2017-2, 161
Complex numbers
  P1986-3, 210
  P2011-8, 25
  S2001-4A, 67

D
Differential equations
  P2012-8, 28
  P2022-2, 59
  S2023-4A, 194
Discrete math
  P2011-1, 23
  P2011-5, 25
  P2015-4, 35

E
Epsilon-delta
  P1968-3, 208
  S2009-6A, 104
  S2010-7, 111
  S2012-7, 126
  S2014-2, 138
Euler's Formula
  P2013-8, 30
Euler's Identity
  S2001-4A, 67

F
Fermat's Little Theorem
  S2005-8, 88

Index

Fibonacci sequence
  P2003-6, 8
  P2009-2, 19
  P2013-6, 29
  P2016-5, 40
  P2023-5, 61
Four Color Theorem
  S2015-6, 149

G
Games
  P2003-2, 7
  P2010-3, 21
  P2011-1, 23
  P2011-5, 25
Geometry
  analytic
    P2001-5, 3
    P2011-2, 23
    P2019-2, 51
    P2019-8, 53
    P2021-4, 57
    P2022-1, 59
  Euclidean
    P2002-6, 6
    P2003-3, 7
    P2004-5, 10
    P2005-2, 11
    P2005-7, 12
    P2006-7, 14
    P2014-5, 32
    P2015-3, 35
    P2016-3, 39
    P2017-1, 45
    P2018-6, 50
    P2023-6, 62
Graph theory
  P2013-8, 30
  P2015-6, 36
Green's Theorem
  S2012-8, 128

H
Heine-Borel Theorem
  S2013-2, 132

I
Induction
  S2003-2, 76
  S2003-6, 79
  S2004-4, 82
  S2010-3, 109
  S2011-6, 117
  S2012-6, 125
  S2013-1, 131
  S2013-6A, 133
  S2014-2, 138
  S2015-1, 145
  S2015-5, 148
  S2016-1, 151
  S2016-2A, 151
  S2016-2B, 152
  S2016-5, 154
  S2017-6, 164
  S2019-5, 177
  S2023-5A, 195
Inequalities
  arithmetic-geometric mean
    S2014-3B, 139
    S2014-3C, 140
  rearrangement
    S2021-5, 182
  triangle
    S2011-6, 117
Intermediate Value Theorem
  S2006-4, 92
  S2014-4, 140
  S2022-7, 189

L
Law of Cosines
  S2003-3, 76
  S2006-7, 93
  S2018-6B, 169
  S2023-2, 194
Law of Sines
  S2003-3, 76
  S2014-5, 140
  S2023-2, 194
L'Hôpital's Rule
  S2014-1, 138
Linear algebra, *see* Algebra–matrix
Logic puzzle
  P2006-8, 14
  P2015-4, 35
  P2021-8, 58

M
Mean Value Theorem
  S2011-6, 117
  S2015-8, 150
  S2016-6, 156

Monotone Convergence Theorem
  S2013-1, 131
  S2019-6, 177

N
Number theory
  analytic
    P2001-4, 3
    P2011-4, 24
    P2012-6, 27
  elementary
    P2001-1, 3
    P2002-3, 5
    P2005-8, 12
    P2006-6, 14
    P2007-2, 15
    P2007-4, 15
    P2008-6, 18
    P2009-4, 20
    P2009-5, 20
    P2010-2, 21
    P2012-1, 27
    P2012-2, 27
    P2014-8, 33
    P2015-1, 35
    P2016-1, 39
    P2017-4, 46
    P2021-3, 57
    P2021-6, 58
    P2022-5, 60
Numerical analysis
  P1969-6, 209

P
Pigeonhole principle
  S2002-4, 72
  S2013-3, 132
  S2016-7, 156
  S2018-8, 170
Probability
  P2003-2, 7
  P2006-3, 13
  P2007-4, 15
  P2008-2, 17
  P2009-1, 19
  P2010-6, 22
  P2017-3, 46
  P2018-7, 50
  P2019-4, 52
  P2023-7, 62
  S2023-3, 194

R
Rational Root Theorem
  S2001-4A, 67
  S2012-3, 122
Ratio Test
  S2004-2, 81
Real analysis
  P1982-5, 210
  P2002-1, 5
  P2009-6, 20
  P2010-7, 22
  P2012-6, 27
  P2013-7, 30
  P2014-4, 32
  P2015-5, 36
  P2017-8, 47
  P2018-2, 49
  P2018-3, 49
  P2019-1, 51
  P2022-7, 60
Rolle's Theorem
  S2001-2, 66
  S2015-5, 148

S
Set theory
  P2004-1, 9
  P2009-7, 20
  P2011-7, 25
  P2014-6, 32
  P2022-8, 60
Social science
  P2002-4, 5
  P2003-2, 7
  P2014-6, 32
  P2016-8, 41
  P2018-8, 50
  P2019-3, 51
  S2015-2B, 146
Squeeze Theorem
  P1982-5, 210
  S2005-3, 86
  S2015-5, 148

T
Telescoping sum
  S2004-2, 81
  S2006-2, 91
  S2008-5, 100
  S2011-5, 115
  S2018-2, 167
  S2018-5, 168

Topology
    P2013-2, 29
    P2016-4, 39
Triangular numbers
    S2001-1, 65
    S2009-7A, 106
    S2017-5, 163
    S2022-5, 189
    S2022-8, 190
Trigonometry
    P2008-3, 17
    P2018-1, 49
    P2023-2, 61
    S2001-4B, 67

    S2003-3B, 78

V
Vectors
    S2001-5, 68
    S2014-3A, 138
    S2021-4, 182

W
Wilson's Theorem
    S2009-5, 104
    S2010-2, 109

The manufacturer's authorised representative in the EU is Springer Nature Customer Service Centre GmbH, Europaplatz 3, 69115 Heidelberg, Germany. If you have any concerns regarding our products, please contact ProductSafety@springernature.com

Printed and bound by CPI Group (UK) Ltd, Croydon, CR0 4YY

26/03/2026

02078965-0001